REVISUALIZING ROBOTICS

NEW DNA FOR SURVIVING A WORLD OF CHEAP LABOR

Steven Baard Skaar. Ph.D.
Guillermo Del Castillo. Ph.D.

DNA Press™

NEW AGE

First Indian Sub-Continent Edition : 2007
by New Age International (P) Ltd., Publishers

Originally Published by *DNA Press, U.S.A.*

Printed in India at Nisha Enterprises, Delhi.

1 2 3 4 5 6 7 8 9 10

C-07-06-1628

ISBN (13) : 978-81-224-2134-7

ISBN (10) : 81-224-2134-2

Original Edition Price: $ 29.95

Indian Edition Price: 495.00

Publishing for one world

NEW AGE INTERNATIONAL (P) LIMITED, PUBLISHERS
(formerly Wiley Eastern Limited)
4835/24, Ansari Road, Daryaganj, New Delhi - 110002
Visit us at **www.newagepublishers.com**

SUMMARY OF CONTENTS

Foreword

What is it, most essentially, about human participation in production - hands-on, physical participation – that seems to be so indispensable for so many industrial, maintenance, farming, warehousing and construction purposes? Answered succinctly, we say "hand-eye coordination". We don't mean to emphasize the athletic sense, as one might apply to a star third baseman in baseball, but rather the broadest sense of everyday activity. Think of "tool-eye coordination" – "monkey-wrench-eye coordination" for instance – or "part-for-assembly-eye coordination" or "pallet-eye coordination" or "paint-scraper-eye coordination" or "fabric-workpiece-eye coordination" or "vacuum-cleaner-eye coordination" or "vehicle-eye coordination" or "lawnmower-eye coordination".

Magnificent as it is, the human mechanism is costlier to employ and generally more feeble than the strictly mechanical part of a system that, for a given category of task, could produce an equivalent or better result. Except for the need to control them using vision, today's mechanisms – with the power, dexterity and/or mobility to handle, with perhaps minor accommodation of the tasks' details, most industrial operations - are both buildable and affordable: Dexterity, in the form of powerful, quick-as-your-eye-can-follow robotic arms, or mobility in the form of a steerable, wheeled platform, become increasingly affordable. If built in large numbers, even a mobile-platform/onboard-arm combination with the power, speed, and freedom-of-movement to field like the star third baseman might be comparatively cheap.

After all, such a mechanism doesn't have to nourish and cleanse millions of cells simultaneously and continuously, whether it's working or not. It doesn't have to be a replica of progenitors capable of surviving and procreating in the wilderness. It need not be able to chew or to spit. It just has to move, catch, throw, tag and snag within the pretty level, pretty uniformly constructed and maintained, playing field of a baseball park. And you can plug it in between innings.

The resistance of the imagination to such an athlete-mechanism isn't so much the strictly mechanical attributes. It's not a stretch, for example, to imagine a very small, wheeled platform motorized to be able to scoot several feet in any required direction in the time it takes a struck ball to reach its playing depth in the field. Nor is it unreasonable that a lightweight, multi-link arm so transported could move itself simultaneously into a position and orientation such that a soft pocket on its end member makes the catch.

What does defy the imagination is successful control of this system: Fleeting early glimpses of the impacted ball must translate instantaneously into early motion to "get a jump" on the ball. Ongoing refinement of such estimates as the viewed ball proceeds must lead to correction of all the system's motors' responses as required to unite ball with intercepting pocket in time and space. Moreover, the combination of illumination and shadow incident on the ball, minute fractions of which would be reflected into the human fielder's eyes, are unique to the moment. Unique to the moment too are the smudges and spin on the ball that affect this reflection of light into the visual sensors, as well as the particular visual backdrop against which the ball must be distinguished.

The wheeled infielder conveys two premises of the book: First, insofar as capability required for the strictly physical aspects of most production tasks (i.e. postioning resolution, speed, versatility of movement, dexterity, mobility, strength, durability), suitable purpose built mechanisms would be inexpensive relative to the cost of employing human labor. But second, this prospective utility and cost advantage cannot generally be realized unless and until new competency of device control comes about. Specifically required is the ability to control using information from light reflected off the surfaces of interest.

Much of the wealth of the industrialized world is a consequence of having extensively built and used powered devices that, each in its own way, achieve ends formerly achieved by human or animal power: steamboats, forklifts, power drills, backhoes, crop dusters, chain saws, jackhammers, tractors, weed-eaters, automobiles, cranes, spray painters, vacuum cleaners and so on. Major accommodations to the design of each of these, however, were specifically required such that their physical guidance and control pass through the remarkable, teachable apparatus of the human eye and brain. A comparable realization of new wealth will attend the exploitation of truly autonomous visual control of means of production as outlined, explained, illustrated, and developed for readers to test, herein.

A robotics revolution has long been expected. But it has been surpassed by the *information revolution* beginning late in the twentieth century, coming into full swing in the mid-nineties with the rise of the internet. The excesses of the almost inevitable "dot-com" bust did not prevent what was already set in motion. Today, the internet has permeated everyday life bringing significant changes in the way we interact with each other and our world.

But what about the *robotics revolution*, a revolution that in the early 1980s seemed as likely as the information revolution; what was it supposed to entail?

As early as the 1960´s there is talk about a robotics revolution. The future would look like an episode of *The Jetsons*[1], or, to use a more contemporary reference, at least like an episode of *Futurama*[2]. Robots would do all menial jobs, operate completely workerless factories, do dangerous tasks like disposing of radioactive waste and disabling land mines.

---

1 *The Jetsons* is an animated television series produced by Hanna-Barbera Productions from 1962 to 1963. Similar in tone to the more popular *The Flintstones*, it also followed the model of a projecting contemporary American life into another time period. *The Flintstones* lived in a prehistoric world filled with machines powered by dinosaurs, while the Jetsons existed in a utopian future of spacecrafts, aliens and gadgets. Its view was greatly influenced by the optimism prevalent in science fiction written in the fifties. This particular future featured robots as a part of everyday life, doing chores that humans would find tiresome or boring, from assembly work to house chores. *(G.D.)*

We are now in the early 21th century, and Arthur C. Clarke´s 2001 prediction of robot use and space travel seems to have been too optimistic by far. The robotics technologies principally used today are essentially the same as those used fifty years ago. Touted early successes, impressive as they have seemed, have proven to be technological dead-ends – fatally flawed. It is disheartening to realize that most robots in operation today are tele-operated by a 'human in the loop" or, more likely, make use of the venerable teach-repeat technique. There has not been - to use the parlance of the nineties - a *paradigm shift*, a true revolution in our lifestyles. The *robotics revolution* has not yet come to pass, certainly in no way comparable to the information revolution.

What have the limiting factors or stumbling blocks to the widespread use of robots been?

Robot systems can be separated into two parts, a software and a hardware component. By hardware we understand the physical body of the robot: its mechanical parts - motors, links, wheels, gears - and its sensing devices: ultrasound transducers, cameras, encoders, laser range-finders, bumper sensors, LIDAR, etc. By software we mean a programmed version of the control philosophy: Teach-Repeat, Visual Servoing, Simultaneous Localization and Mapping (SLAM), Behavior-Based, Calibration, Tele-Operation, etc.

The authors believe that the robot hardware available today is essentially able to do a very large fraction of useful tasks now requiring human labor, and much of what is shown in sci-fi fantasies. The big stumbling block has been the software. The problem is not the software itself, but the control philosophies developed so far. Perception of an environment, in particular human-like visual interpretation, has been notoriously difficult to understand or replicate, with evidence suggesting that replication may not be feasible.

The New DNA we propose is a way to re-visualize the field of robotics: With the technology available today, we cannot have a *Star Wars*-level of robot autonomy, but we might achieve a high level of usefulness, enough to make robots part of everyday life. Robots would still need human guidance – the parts related to perception - but they would be truly autonomous while executing a task. Some of the techniques discussed herein are extensions of old paradigms, like teach-repeat, and others are a combination of the strengths of existing paradigms, like Camera Space Manipulation.

---

2 *Futurama* is an American animated television series created by Matt Groening and David X. Cohen. The show aired in the Fox Network from March 28, 1999 to August 10, 2003. The characters had the signature look that Groening popularized in his other creation, *The Simpsons*. Set in "New New York City" in the 31st century, the chosen time frame allowed the writers to introduce ideas from many varied sources, in particular, science fiction from mid to late twentieth century. One very interesting idea that sadly, could not be developed further (due to Fox canceling the show on its fifth season) was a parallel society of robots. This robotic underclass was a combination of the utilitarian robots displayed in shows like The Jetsons with their more malevolent counterparts that were a fixture of American pop culture in the eighties and nineties. Futurama airs on several syndicated cable networks. *(G.D.)*

One likely prospect discussed in the later chapters is a factory where workers do not need to be physically onsite. They instead provide the perception for, and high-level direction of, the robots from any part of the world, using the internet, for instance.

All of the interrelated forms and paradigms of the book have enabling elements that are nonintersecting with mainstream robotics, although the history of mainstream robotics is used for context and to illustrate the subtlety of many of the issues involved.

The paradigms draw pragmatically on the remarkable human ability; however, there is no effort to essentially understand or replicate this ability, but rather to complement it. In no case, however, is this complementary human role one of ongoing device guidance and control; the systems are truly autonomous while running.

Steven Baard Skaar, Ph.D.
Guillermo Delcastillo, Ph.D.

CHAPTER 1

WHERE ARE THE ROBOTS?

Theories aside, their "widespread absence" speaks loudly; but what is it saying? One main purpose of the present volume is to make understandable the so-far prohibitive difficulties of manifesting the long-expected proliferation of autonomous mechanical assistants and workers. This chapter begins the discussion.

Early in the 1990s a former graduate student of the writer's home department returned for a visit to Notre Dame. He was looking for engineering ideas to help his current employer, a maker of commercial floor-maintenance equipment, out of a nervous bind. Japanese competitors, they believed, would soon produce and market autonomous units. Rather than relying on continuous human guidance, these new scrubbers and vacuum cleaners would cover the required floor space autonomously, completely on their own. No human labor. No human error. No overtime pay. And while the cost of such units would undoubtedly initially exceed that of conventional equipment, customers' labor savings would certainly result in a large shift in market share in favor of the Japanese if the challenge went unanswered.

As discussed in Chapter 10, this intriguing problem did actually lead to some thinking and eventual development at Notre Dame University, but only gradually - not in time to mitigate the urgent concern. So the company moved ahead internally, engaging talented young engineers with a can-do attitude to create a prototype to meet the seemingly certain global challenge. Within a couple of years the writer visited the company to witness first-hand the fruits of this effort. A large unit, perhaps four feet tall, two feet wide and five feet long, an adaptation in fact of a previously existing, human-guided floor scrubber, was released in an empty room. Automatically, impressively, it traversed 360 degrees around the perimeter of the room, sustaining a short distance from the outside walls. When it reached an open door along the room's perimeter it moved past it, undeterred by the interruption in the tracked wall. When one of the company's demonstrators stepped into its path it came to rest, waiting for the impediment to move, at which time it resumed action without a fuss. After completing a full outermost pass, it extended its distance to the room's wall to cover a new swath of floor space. In continuing this way it spiraled in to the center, exhaustively, with the exception of room's corners and extreme edges, covering the floor using no human guidance at all.

Although the room was simple in geometry, basically a rectangle, and "obstacles" within it always eventually stepped out of the autonomous unit's way even as it waited, this was, in the mindset of those in attendance, a success. The prototype was undoubtedly limited, but the design it was thought would find early adopters, customers whose needs were compatible with the demonstrated capability; and indeed it did.

More to the point of this optimism: the system was of course based upon newly written software; and by this time the *modus operandi* of successful software development was well-known and unimaginably successful: Gain an early presence in the marketplace. Rely upon users' ability to cope with imperfections; in fact, expect early adopters' eagerness to participate in a helpful product-development feedback cycle due to their characteristic fascination with high-tech novelty. And simply use "version 1.0" as a base, a building block off of which all manner of subsequent competent complexity would follow. It had also become clear, moreover, that the costs and capabilities of the system's constituent electronic parts had dropped impressively over the previous decade, even as capabilities grew.

Time, it seemed, was on the side of the earliest entrants into the marketplace. The question of intrinsic real-world feasibility of autonomous floor maintenance may not even have been entertained seriously. After all, autonomous aircraft could take off, land and fly missions without a human pilot. Computers could often beat grand masters at chess. Why should the custodial class be so indispensable?

Similar optimism based on early, partial successes penetrated academia. One example is autonomous road vehicles. As far back as the 1980s television broadcasts showed footage of vehicles driving autonomously within highway lanes. This helped fuel the perception that it was robots, not particularly the internet (what was the *internet*?) or even information exchange generally that was the great hope of burgeoning, often-spoken-of/vaguely-understood "high technology". A distinguished German academic, Ernst D. Dickmanns, then at Universitaet der Bundeswehr, had made particularly impressive progress, among other things applying computer vision to follow less-improved roads - roads that did not have painted lines to demark lanes. The present writer was tasked by organizers with making a written record of a 1997 panel discussion where the Professor's remarks surprised fellow participants at an international workshop held on Rhode Island's Block Island: With computers two hundred times faster than those currently applicable, he opined, we should be capable of creating autonomous vehicles with a competency of driving on typical roads in Germany about equal to the average human driver on the same roads. Know-how for the needed algorithms was, one inferred from his remarks, already extant; it was just a matter of computer speed catching up. (Moore's law, which states roughly that computing power doubles every eighteen months, would put the date of arrival of such a system just a few years from now.)

In addition to the commercial sector and academe, the U.S. government and governments in Europe and Asia have also been caught up in the optimism. Military and space missions with their attendant dangers seemed like natural applications of autonomous systems. EVA or extra-vehicular activity on the part of astronauts, for example, is expensive and dangerous. On-orbit servicing of the Hubble telescope by astronauts, while achieved successfully in the past, may presently be deemed unduly risky. So while EVA has been used and continues on the space station, at one point it was announced that robotic servicing is the likely resort for Hubble. Similarly, exploration and development of lunar or planet surfaces, if conducted robotically, might reduce the cost of missions in the short term by factors of ten or more. At the same time, military use of autonomous air vehicles long ago began paying dividends in terms of utility, cost and safety. Why not extend this concept to military ground operations?

Government's military and space initiatives in this field seemed to carry with them an appealing political bonus: the likelihood of "dual use", the extension of space/military technologies to civilian and commercial application. Just as the federal government had played an instrumental role in private-sector adoption and proliferation of color and high-definition television as well as the internet and solid-state computing, perhaps it could jump-start robot use.

The Department of Veterans Affairs (previously named the Veterans Administration) was an early player in this cause, with an early-90s introduction of an autonomous delivery robot into certain of its hospitals. A way to defray the labor cost of delivering items such as medicine within hospitals, early prototypes worked well most of the time. As with the floor-maintenance system discussed above, the Hobbit-sized machines could tolerate open doors as they tracked hospital-corridor walls; and they would stop when interfered with, accidentally or on purpose, by pedestrians. Some would 'wow' an interfering corridor walker, in fact, with startling spoken politeness. This kind of pleasing responsiveness added to the optimism. Future, similar layers of programmed-in contingent responses evolved and perfected over a long enough period of time would, it was thought, continue. Each in its turn might rectify problems with the system's response as new circumstances and challenges were encountered. First, a kind of passable proficiency for early adopters would be reached; and later, if the precedents of word-processing and spreadsheet software held sway, these systems would get so good as to become necessities. If version 3.2 didn't get them to buy it in Peoria, version 5.6 surely would. In the contingency provisions' very complexity, however, overall behavior became unpredictable as programmed responses interacted in unforeseen ways with an infinitely variable, sometimes willful environment. The better historical comparison was not word processing or spreadsheets but the frustration of Apple's Newton Pad's capability for reading natural handwriting.

But if such aspirations eventually, inconspicuously, drifted out of sight and mind, the canary that saliently alerted the public to stifling government futility occurred with the U.S. Defense Advanced Research Projects Agency (DARPA) Grand Challenge. First held in March 2004 this was a competition for fully autonomous vehicles to compete on an under-300 mile, off-road course in the Mojave Desert. Maybe, was the implicit concession, "robotics" isn't after all the kind of science that demonstrably yields fruit with academic and government-laboratory attention, blue-ribbon panels, conference proceedings and the like – self-correcting, and an overall sound investment for taxpayers – at least, perhaps, not in the way that it had been pursued the previous twenty five years. Maybe some paradigm shifts were in order; and maybe the greater public should be allowed the opportunity to demonstrate them.

As it happened none of the entrants got more than a few miles in the first year of the Grand Challenge. Its object, in comparison with myriad other tasks attempted with less publicity, may not be especially helpful in the real world if later some do. Rough point-to-point autonomous navigation is relatively easily accomplished in the air for instance, and on the ground it is highly dependent upon the particulars of the course and circumstances. The very scale and spectacle of the Challenge, however, was seminal in that it prompted discussion, much of it on the internet, with broad participation as to the intrinsic impediment. Just what do humans bring to these tasks of ground transport and navigation, or for that matter road-vehicle driving, cleaning, etc. that is so very elusive in artificial systems?

Something of a common wisdom emerged regarding the missing ingredient. It had perhaps best been explained some years earlier by MIT's Professor Rodney Brooks in an episode of the television show "Secret, Strange and True". Written by Lawrence Sanfilippo, 'The Battle of the Robots' episode documents three ambitious intelligent-robot developments. Two were financed with government funding (EU and US), one with personal resources. None was to succeed. When asked why researcher Brian Scassellati, then an MIT graduate student, had not reached the goal of making a robot see something, decide whether it's alive, and learn from it, Brooks placed two objects on a surface before the documentary's camera. One was a pair of glasses, the other a cell phone. He explained that even after years of related efforts, there was still no general-purpose computer program able to use the information in a digital image to identify the two objects.

It seemed startling if you thought about it. Most any digital-TV viewer of the episode would recognize these things instantly, accurately and without doubt. Any viewer would know whether the clamshell phone and eyeglasses were folded or unfolded, and in what way, approximately, they were juxtaposed. The viewer's correct instantaneous recognition would not depend upon color or other reflective characteristics of the objects' surfaces or of the surfaces upon which the objects were laid; no need for any special lighting; background clutter in the laboratory wouldn't interfere. If asked to pinpoint a hinge of the eyeglass rim, any viewer could locate that juncture on the image immediately. Even if the rim were partially occluded one could similarly, easily trace a finger along the ear-curvature portion as it appeared in an image. These human abilities seem remarkable only when contrasted with the best computer vision has to offer. And it is the exploitation of these human abilities using "point and click camera space manipulation", one of the new technologies presented herein, that can multiply human productivity many-fold.

But if the state of the image-analysis, object-recognition art is ineffectual with two separated, distinct and common objects in a laboratory with more or less constant artificial illumination, what of the autonomous desert traveler, or road vehicle, or vacuum cleaner, or medicine deliverer? Is that narrow patch ahead rotting cactus or the bottom of a ditch? Is the object nearing the vehicle an oncoming bus or the shadow cast by a flock of geese? Is that vertical line in the digital image the edge of a doorway or a decorator's latest wall embellishment? Is the bright object on the floor a sheet of paper or the reflection of sun shining through a windowpane? While there are certainly, for any given dilemma posed, disambiguation alternatives to general-purpose vision (the kind we take for granted in ourselves) that could be devised by a clever engineer, the question is how usefully a system without our more general visual capability can be made to work in the real, infinitely variable world? So far the answer seems to be "not very". As discussed below, in fact, about as much today as ever, the most used systems that fall under the broader definitions of robots don't seek to replace effective human vision at all, but rather to exploit it and to complement it, albeit in limited and limiting ways. A big part of the "revisualizing robotics" of the present volume actually entails *new ways* - "point-and-click camera-space manipulation" mentioned above and "teach-repeat extended to nonholonomic (wheeled) robots" - for *further* leveraging the human visual recognition/apprehension capability. These we believe will produce utility and application that, while not Spielberg autonomous, should multiply the number of units of today's almost cheap mechanical dexterity and/or wheeled mobility that are profitably useful and significantly autonomous.

- *First of two ways robots today rely upon human visual guidance: Teach-Repeat*

Continuing from its beginning to now, the predominant real-world robot-control category, what we here call "teach-repeat", only takes advantage of the "dexterity" (or kinematic versatility - lots of joints, judiciously arranged to permit needed pose realization of the manipulated body) of the robot to the extent that it allows a human "teacher" to imprint upon the system a very specific set of poses. Discussed at length in Chapter 2, this is what enables most of the robots professionals in factories deal with. It is based upon a simple premise: An articulated mechanism (e.g. an arm with shoulder, elbow, wrist joints) will return an object within its grasp to a "taught" position and orientation in space if the internal rotations of these joints are returned to their counterpart rotations as "taught". It's hard to verify this with our own arm because we can't naturally return say an elbow to any given number of degrees of extension. But for a robot it's easy both to measure such a rotation and to coax it one way or the other to achieve a precise target angle. See *Figure 1-1*. This intrinsic advantage of

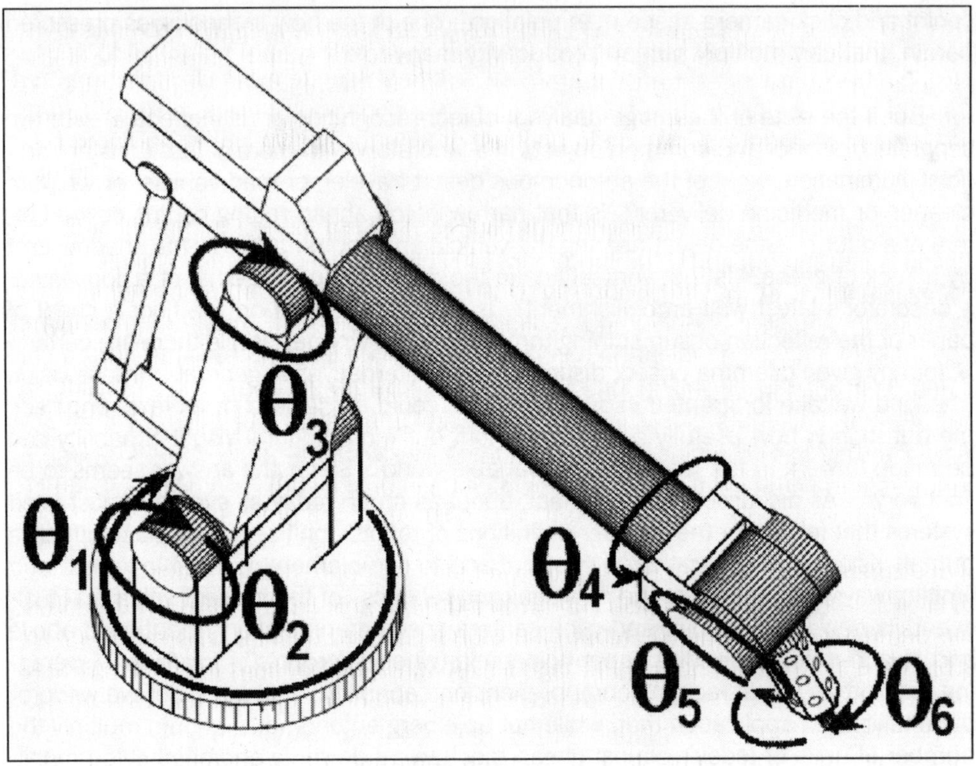

Figure 1-1. Robots are able to drive each of, in this case, six joint angles to any prescribed number of degrees.

machines together with many others such as steadiness and strength make it all the more remarkable that our visual ability results in the human factor winning out over dexterous machines even in most factory applications where repetition is inherent. The main reason for our advantage over the teach-repeat machines lies in a subtlety: It is costly to repetitively place each new workpiece on an assembly line in exactly the same position and orientation in space - copy after copy - as was the prototype work-piece during teaching. Teach-repeat robots require this precise, active pre-placement of the current workpiece. Before a spot-welding robot, for example, can fix a new juncture in a car's frame, elaborate equipment must previously have brought the pertinent pieces to a specific location in space, a location identical to that with which the robot spot-weld pose was taught. So the automobile spot-welding robot merely drives each of its internal joint angles to its taught number of degrees of rotation. People, thanks to our eyes and "hand-eye coordination", don't require such costly fixturing; we can and do respond to "as located". We see and then we reach and then, continuing to monitor progress with our eyes, we weld, or assemble or paint or rivet - or thread a needle. We're really incapable of doing otherwise. And, where robotic teach-repeat is called for, it is people who accomplish the teaching of the robot in the first place, most often using the visual capability as discussed in Chapter 2.

But when it comes to autonomous *indoor-vehicle navigation* such as might be useful to a blind rider of a power wheelchair – or perhaps as the control basis for an autonomous floor scrubber – robot-trajectory repetition wouldn't carry with it the cost of repositioning objects in absolute space. That's the message of Chapter 10. A wheelchair or vacuum cleaner may today be called upon to repeat the same sequence of positions within the plane of the floor as was used yesterday without a costly need, prior to each repetition of the taught move, to "reposition the walls" or "reposition the bathroom fixtures". There is only the *passive* need to keep these things where they were when the teaching occurred: If yesterday's action accommodated furnishings, bathroom fixtures and walls, today's will too.

Actually *realizing* teach-repeat is not so straightforward for the vehicle, however. Merely returning internal angles to their taught poses is not enough because the key geometric property that makes teach-repeat so very easy for the typical industrial robotic arm, "kinematic holonomy", is not shared with wheeled vehicles. *Both* "holonomic" industrial arms *and* "nonholonomic" wheeled vehicles get the positioning job done using motor-driven rotation, but in the case of the nonholonomic vehicle, recovery of a taught, physical location and orientation in space is not guaranteed by returning each wheel to a counterpart, taught rotation (Chapter 9). More is needed; and the enabling additional provision for achieving teach-repeat with a vehicle is another "new-DNA" technology presented herein. We argue in Chapter 10 that paying the price to accommodate this extra need of "extending teach-repeat to nonholonomic robots" is well worth the engineering effort due to the wide range of real-world-practical, real-world-useful autonomy it can deliver. Blind "teach-repeat" is responsible for control of far and away most robotic arms sold today. But the burden it carries of active, precise introduction of each new workpiece severely limits the number of units sold and used. The *passive* counterpart requirements for use of teach-repeat with nonholonomic, wheeled vehicles, and the commonness of its applications, however, will, we believe, result it this manifestation finding orders of magnitude more individual expressions. Adding to this optimism, we find that coping autonomously with newly introduced obstacles – generally not an issue with teach-repeat for industrial arms – is also enabled with a prior teaching run, as detailed in Chapter 10.

• Second of two ways robots today rely upon human visual guidance: Human in the Loop

When many think of robots they may have seen with some frequency, often battling robots on TV come to mind, or competing robots such as those of the FIRST program. Unlike "teach-repeat", where exploitation of the human's remarkable visual/perceptual ability is one-time, during teaching, "human in the loop" partners continuously and in real time the human visual ability with the ability of mechanisms to project mechanical effort out into the world and space. No less than NASA uses this ongoing, real-time synergy between human visual understanding and the ability of the machine. One of the problems of including human-in-the-loop under the heading of robotics is that there is no definitive distinction between the control of, say, Canadarm by an astronaut in the space shuttle and the control of the dry-wall-engaging arm by the operator in *Figure 1-2.* Continuous, real-time human-in-the-loop control, nonetheless, enables all kinds of devices; some of them may be regarded as robots, others not: road vehicles, forklifts, power wheelchairs, power tools, backhoes, cranes, hobbyists' radio-controlled airplanes and the drywall-loading apparatus of *Figure 1-2* to name just a few. As discussed in Chapter 5, the human ability to serve so effectively in these various roles is, from the point of view of one trying to emulate similar capabilities with artificial vision, quite remarkable. But it does have severe limitations.

These limitations were acute considerations, for example, as NASA's look into the possibility of robotically servicing the aging Hubble telescope developed. Due to safety concerns NASA had considered ruling out using astronauts in EVA, even though this had been accomplished successfully previously. Instead the hope was to apply dex-

Figure 1-2. Human-in-the-loop mechanism control works in this case *if* high speed isn't needed, if visual access is direct and reasonably proximate, and if a large gap affords plenty of tolerance for error.

terous mechanisms docked with the orbiting telescope. As with most tasks, the problem is not identifying hardware able mechanically to complete the task: A robot can definitely do the work, former astronaut John Grunsfeld, NASA's chief scientist, said of the Hubble-servicing challenge. The Canadian robot "Dextre" was one possible choice – but the tricky issue of control left uncertainty.

Grunsfeld has explained that an operator familiar with the telescope - perhaps a former Hubble-servicing astronaut like himself, would control the Hubble-servicing robot from the ground, in real time. He added that even if a robotic mission did not fully succeed, engineers would learn plenty to apply toward future efforts at remote operations on the Moon and Mars. In a series of long-distance tests, NASA astronaut Michael Massimino sat at a control console in Houston and operated a prototype robot in Maryland. The idea was to practice work that would have to be done on an unmanned servicing flight. One of the great attributes of humans as we place ourselves into various mechanism-control loops is the ability to improve with practice.

• *Why, after all these years of robotics research, do we still rely on human-in-the-loop control?*

But if the process of robotic servicing of Hubble cannot be autonomous, and is instead "more like an orbital video game", with added difficulties of indirect visual access and some time delay, is it the *Secret, Strange, and True* problem of automatic object recognition/location within an image that forces a resort to human-in-the-loop control?

Actually, an impediment other than artificial image analysis arises, the subtler difficulty of calibration - a difficulty with prior historical significance in that it had prevented costly factory-automation efforts in the 1980s from succeeding as envisioned. With man-made artifacts of precisely known geometry artificial image-analysis can often be solved reliably. For the eye glasses and the cell phone on a flat surface, for example, while a single image, sufficient for the human, may not have sufficed, multiple images input into a computer, each acquired with different lighting conditions may have made the objects locatable in the image - provided the geometry of the objects had been characterized in a data base in advance. Special structured lighting combined with "image differencing" – subtracting pixels' intensity values in a digital image with just ambient lighting from pixels' image intensities with the structured light turned on – can, especially with a suitably limited range of expected object orientation, yield reliable computer image interpretation and precise object-feature location within a digital picture of, say, a pair of eyeglasses, or a cell phone - or a module-receptacle slot on the Hubble. Chapters 6 and 8 discuss use of one kind of structured lighting, "laser spots," (although stripes are more common.)

In addition to manmade objects such as Hubble-telescope components, a factory assembly line often presents to cameras objects with precisely known geometry and therefore just such an opportunity to locate objects of interest artificially within an image. And automobile manufacturers believed exploitation of this fact would unleash robots' ability to wield their dexterous versatility as each workpiece's actual location required, copy by new-car copy – resulting in full factory automation. The aforementioned "subtler difficulty", however, caused expensive 1980s experiments to make robots respond to workpieces "as-located", famously, to fail.

General Motors then-CEO Roger Smith was an aggressive developer and adopter of new technology. GM invested heavily, as did other automobile manufacturers, in the vision of workerless factories. Hard automation, and even teach-repeat robotics, already in use, could only take this vision so far, however: Teach-repeat's workpiece-prepositioning requirement was not practical for many tasks for which humans were currently employed. Since computer image analysis to locate workpiece junctures of interest in a highly structured factory was feasible, however, visually guiding robots seemed a natural way to manipulate tools, welding heads, or parts to be added to an assembly in response to actual, variable workpiece location.

The dexterous machines themselves, robots capable mechanically of handling the various tasks previously achieved with human dexterity, were not a problem. What did turn out to be insurmountable, however, were the vagaries of calibration – calibration of robots' geometric "kinematics" and calibration of participant vision systems (Chapter 3).

In the late 1980s and early 1990s the writer queried automobile-industry engineers whose specialties centered on assembly-line operations. The 1980s experience and tales of the experience were fresh in their minds. The straw that some recalled had broken the camel's back, the automation task that was to be the go/no-go decision point, turned out to be familiar and mundane: using vision for robotically positioning a new wheel onto a new automobile brake plate. Flat-tire changers will recall this jacked-up, five-threaded-bolt "brake plate" as their own side-of-the-road target for positioning the spare taken from their trunk. On the side of the road, we tend to "feel around" until the thing pops into place; in the factory, the robot would move straight in with the wheel but too often, unceremoniously, just miss, albeit slightly, but with no haptic recourse.

This dichotomy between human vs. robotic assembly motion was cited by former astronaut John Grunsfeld, NASA's chief scientist, in the context of Hubble servicing. From experience, Grunsfeld said, equipment sometimes gets stuck while being swapped out. Astronauts can feel what's going on, stop, and make adjustments. On the other hand, "robots can do really pure motions." But those pure motions must be on target, and for this a calibration approach to autonomous, vision-based robot guidance has proven unreliable, and therefore, in three dimensions, has largely been abandoned.

- *Camera-space manipulation and its relationship to Calibration*

That it is *possible* to complete robustly tight-tolerance assembly tasks such as wheel loading with only image information was the objective of early-90s demonstrations of the alternative strategy herein developed: camera-space manipulation. Three of the videos in **http://www.nd.edu/~sskaar/** illustrate the "wheel-load" objective being achieved autonomously by a six degree of freedom industrial robot using vision alone. The video demonstrations do not apply the contact feedback humans would invariably use to do the same thing. This demonstration of the ability to control Grunsfeld's "pure motion" of a robotic arm robustly and precisely relative to an arbitrarily placed workpiece or object of interest becomes a necessity in the many situations where feeling the engagement of joined pieces is incompatible with the task. The task of *Figure 1-2* is one such example; contact prior to insertion could ruin the drywall panels. The experienced human operator can control the device without relying on contact sensing because the targeted gap, whose thickness, which together with the thickness of the fork dictates the room for error, is relatively large; also, unlike the wheel-load problem, entry visibility doesn't become occluded by the entering object itself; and finally, the absence of the urgency of a moving assembly line means time allowed to realize insertion is not constrained. For painting, other kinds of surface preparation, welding, stacking, etching, cutting, drilling, riveting and precision engaging, the steady, pure motion of robots would be ideal – if only robots' motion could be reliably, robustly, precisely and autonomously controlled with respect to "as-located" workpieces.

Most often calibrated vision for three-dimensional robot control is infeasible on the factory floor, let alone in space; hence the resort to a very difficult human in the loop option for the Hubble. Nevertheless, if calibration *could* be made to yield 3D precision of dexterous-robot control the very difference between it and the myriad human "knacks" for getting the job done could revolutionize the way we construe all kinds of productive tasks. We think of what it means to *do* something in terms of the techniques distilled through history from much trial and error of *human* attempts. The task, the technique, the trade has evolved based upon refinement of best practices vis-à-vis a replicatable human capability. If enough talented individuals with reasonable amounts of practice can bring about a particular set of positioning objectives and ends, we leave it at that: the task, the technique, the trade has been defined. Otherwise, we modify the design of the widget that must be built, or we create a new tool, or we introduce a new reference mark into the operation – anything to work within the de facto constraints of the human organism's (albeit remarkable) abilities.

One thought experiment from carpentry: Suppose an extraordinary new human being comes onto the scene. This fellow is tasked with using a circular saw to cut a groove at a 45 degree angle relative to the board as shown in *Figure 1-3.* The depth of the groove must be precisely half the width of the board. And the thickness of the groove must be exactly two blade widths.

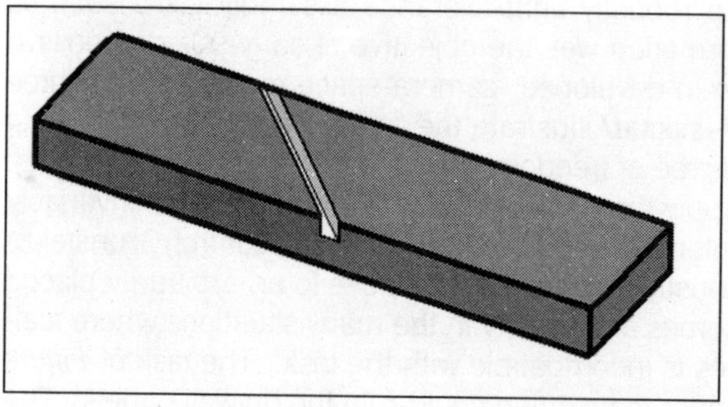

Figure 1-3. A carpenter able to produce a perfect groove with just a spinning blade at the end of his arm having direct contact with the arbitrarly placed piece, would not need the time-honored tricks of the trade.

Now suppose that, rather than adjusting the blade depth on the saw to half the board's width, and rather than penciling in a forty-five-degree angle onto the board's surface, the worker merely walks up to the blank board and creates a motion with his arm that is straight and pure and exact. Other than the cutting blade itself, no part of the saw rests upon or is in contact with the board. Two parallel strokes of the rotating blade at the end of his arm and the result is perfect. Best practices of the trade would change indeed if this fellow could be cloned.

Now a dexterous robot has all of the mechanical ability needed to do just that. If calibration worked perfectly as envisioned, and image analysis were not an issue (since the blank board has geometry known apriori), the motion could be created reliably in just the same way.

What, in the context of visual robot control, is this potentially revolutionary "calibration" and why did its vagaries preclude realization of the large and determined plans of the automobile industry in the 1980s? Actually, calibration has two parts: first is calibration of the cameras, and second, calibration of the robot mechanism. As mentioned above, for an articulated robotic arm, driving each of the joints to a stipulated internal angle is straightforward; it is essentially "what robots do", and it is the sine qua non of "teach-repeat" as it is practiced today (Chapter 2). With calibrated vision, however, cameras measure the coordinates or location - position and orientation - of a given "workpiece", or object of the robot's current maneuver. Computers then do a very rapid calculation based upon the geometry of each of the robot's various "links" (think for example of the length of your forearm), as to just what all the internal angles need to be driven to in order to place the robot's load as required with respect to the workpiece. This is why the successfully calibrated system could perform like the imaginary carpenter, the one able to cut grooves of a precise depth, angle and location relative to an arbitrarily placed board. Whereas human beings are good at bringing about "closure", or a meeting of those physical junctures that must come into contact in order for the task to be achieved, a calibrated robot and vision system would be master of the entire, quantitative geometry of its own spatial domain. It could establish, hold, or move through any required physical juxtaposition - position and orientation – between tool and workpiece as quantitatively specified for the maneuver. It does so by calculating the six angles, θ_1, θ_2, θ_3, indicated in *Figure 1-4*, and the wrist angles θ_4, θ_5, θ_6, not shown. These are found for each juncture of the two passes of the spinning blade.

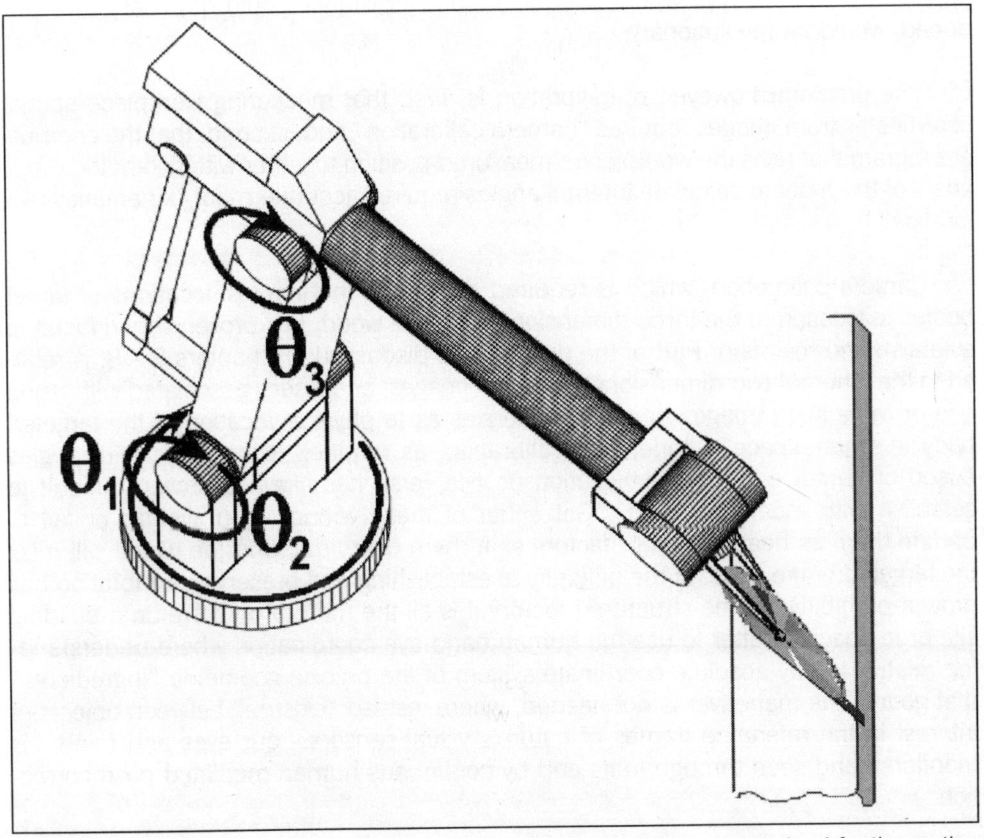

Figure 1-4. The first three of six internal robot angles needed to be determined for the motion sequence to create the slot are indicated. They change with each juncture of the motion. With calibration all angles are calculated precisely in advance of the motion using image information from calibrated cameras together with a prior calibration of the robot's geometry or "kinematics".

So provided image analysis or understanding is achieved, automatically or with some human help, calibration brings to the table brand new prospect, unlike anything humans would be capable of. The calibrated systems' mastering of its domain means that every surface juncture of the workpiece has precisely known physical location relative to every other surface juncture. And the calibrated robot is capable of delivering any selected juncture on the manipulated end effector to one or a sequence of precise, mathematically prescribed positions and orientations with respect to any selected point or set of points on the target body. This is what "CNC machines" are supposed to do; but their domain and application is highly restricted relative to the prospect for robots. CNC machines' absence of vision means that fixturing and geometric characterization of the workpiece are essential prerequisites. Calibrated, visually guided robotic systems, if they truly worked as conceived across the range of applications and settings where portable robotic assistance could physically be introduced, would be revolutionary.

The problem, however, of calibration is, first, that measuring workpiece spatial coordinates from images requires "camera calibration" and, second, that the computer program that uses the workpiece's measured position together with geometric properties of the robot to calculate internal angles requires accurate robot "kinematics calibration".

Camera calibration, which is required to convert the images' locations of target bodies to location in the three-dimensional physical world, has proven very difficult to establish and maintain. Part of the difficulty, as discussed in Chapters 5 – 8, is related to the inherent two-dimensional nature of images, and issues associated with using one or more such images to make inferences as to physical location of the targeted body in "three-space". Kinematic calibration, as required to compute joint angles based on target end-member location or vice-versa has likewise proven difficult to establish with much precision. Get either of these wrong, even slightly, or fail to update them as heat and other factors shift them over time, and the wheel will miss the targeted brake plate. If the difficulty of establishing and preserving calibration has proven prohibitive at the structured factory it is all the more prohibitive on a building site or in space. Better to use the human hand-eye coordination where understanding relative to any absolute coordinate system of the precise geometric "ingredients" that go into the maneuver is not needed, where instead "closure" between objects of interest in the reference frames of nature's visual sensors - our eyes and mind - is monitored and seen through to its end by continuous human-mediated pose correction.

NASA, therefore, sees fit to use "human in the loop" in space; and factories use it on the ground. The problems of calibration do not apply with human-in-the-loop control because the maneuver is pursued and realized in the reference frame of the organism's visual sensor. Sometimes, with both industry and NASA, the "mechanism" that the humans control is their own, natural one (i.e. body and limbs – EVA, for instance). Increasingly, however, factories find ways to let human visual guidance operate machines with the motion versatility needed to complete the task at hand in order to reduce repetitive-motion injury. The nearby presence of the guiding human eye in the factory actually makes this quite a bit easier than in space – especially, with the latter, if the guiding human is far removed and on the ground. With Hubble servicing, it may seem that there is no real alternative.

But couldn't nature's paradigm of bringing about closure or accurate relative positioning of objects of interest in the desired way be achieved in the reference frames of *artificial* visual sensors - cameras - just as well as in the natural visual reference frames of humans? Couldn't the calibration issues thereby be obviated? If humans position well without any explicitly numerical knowledge, perhaps dexterous machines could do just as well or, using their various mechanical and "number-crunching" advantages, even better than humans by executing maneuvers in the reference frame(s) of the visual sensor(s).

In fact, that is the premise of "visual servoing", an object of great academic attention over the course of the last twenty years or so (Chapter 4). It is also the premise of the concurrently developed "camera-space manipulation" a different paradigm, one that has elements of both visual servoing and calibration, and a big part of the "new DNA".

A rough understanding of the distinction between visual servoing and camera-space manipulation can be seen by returning to the thought experiment of the carpenter making the diagonal passes of *Figure 1-3.* Visual servoing would be similar to an ordinary human carpenter, one whose hand-eye coordination would be more or less limited to establishing and preserving closure of the blade vis-à-vis a guide line previously placed onto the board. Visual servoing relies upon the idea of using feedback in order to bring together two junctures in the visual reference frame. Camera-space manipulation *would* have most of the advantages of the imagined carpenter, that is the advantages calibration would have if it could be established and maintained. In the board problem, there would be no need to draw a reference line and all six position and orientation components of the saw's relationship to the board would be precisely controlled.

In practice camera-space manipulation has properties that make it practically useful for achieving precision never possible with even the most careful global calibration, including the ability to refine camera-space kinematics estimates almost to an unlimited degree of precision as cameras are allowed to zoom in to a small region of interest, and as multiply redundant visual and joint-rotation samples are applied to estimates of the corresponding camera-space kinematics and ensuing motion of the mechanical arm nears concomitant perfection. Additionally, CSM retains the robustness and reliability of doing what humans do – construing and pursuing the maneuver in the reference frame of the visual sensors.

Joining the best of all worlds is our objective: control in the sensors' reference frame, using estimation based upon quantitative readings, both visual and of joint rotations; applying human vision to establish via point and click task objectives, or, where key geometric knowledge of workpieces allows, image differencing and structured light to do so automatically; and applying the pure steady motion of artificial dexterity. These CSM strategies join in this volume the aforementioned extension of teach-repeat to nonholonomic, wheeled robots for autonomous indoor navigation, and "mobile camera space manipulation" for applying the mobility and added reach of a wheeled base with onboard mechanical dexterity to realize the robust relative positioning of camera-space manipulation with the range of a mobile base. Together, future prospects of this new DNA are visualized in the final chapter. The authors strongly suspect that these future prospects will tend to be underestimated for one particular reason. When thinking of "replacing" human dexterous and mobility effort, it is natural to do so in terms of those actions and sequences that have evolved over the millennia so as to be optimized vis-à-vis the set of abilities with which we are born. As mentioned, accommodations to these abilities in the way tasks are, in our experience, sequenced and pursued in manufacturing and other productive effort may be unneeded and unadvisable when the pure dexterous motion of a robot under the kind of visual control envisioned becomes available. *Figure 1-4* represents just one illustrative example where there is no real human counterpart ability to the indicated action, and where the visually guided machine's ability obviates all kinds of tradesmen's best practices. A second example, one involving actual experimental results, in another woodworking task, hole drilling, is presented in Chapter 8. Yet the inability to intuit such replacements of best practices is likely to result in an underestimation of efficiencies which may result from new best practices – analogous to ways that a motivated population learned to use, say, spreadsheet software – that may lie ahead.

CHAPTER 2

"TEACH AND REPEAT"

THE CURRENT WAY INDUSTRY CONTROLS ITS 3D STATIONARY, HOLONOMIC ROBOTS: SIMPLICITY, RELIABILITY, BUT SERIOUS LIMITATION.

How do those robots actually in use work? Simple to convey and understand, the limitations associated with this far-and-away most common mode of operation are key to appreciating impediments to more widespread robot use.

Think about the automatic-seat-position memory in your car. If your car doesn't have seat memory, fantasize. You open the door; your seat adjusts - automatically - to its most open, most inviting setting for an entering driver. Then you get in and turn the key. The seat moves again, positioning itself to perfectly accommodate your particular proportions and driving preferences: perfect arm distance to the wheel; perfect leg distance to the pedals; angle of back inclination just the way you like it.

How does this turn-of-a-key bliss come about? Teach-repeat. And who is the teacher? It is you of course.

Some BMW cars have driver seats with eight different motors, each of which can rotate to adjust some aspect of the seat "pose" - or position and orientation. One motor might control seat-back angle; another lumbar-support intrusion; a third, elevation; and so on. What is important about each of these motors is that each is part of a "position servomechanism". Such a device is capable of applying the electrical power needed to rotate the motor up to but not beyond a certain, predetermined, "taught" number of degrees. When you adjusted the seat-back tilt with a switch on the side of the seat, and pushed the memory button, you taught that motor/servomechanism an angle it would remember until re-taught. The same goes for each of the other seven motors on your BMW car driver's seat (remember, this is fantasy).

And the same goes for the overwhelming majority of robotic arms in use today. Rather than accessing position buttons on the side of the seat as in *Figure 2-1*, however, a "teach pendant" is used, a panel with buttons that allow the human teacher to "jog" or move incrementally at will each robot axis. In the case of the six-axis robot of *Figure 2-2*, this would control six different motors, the very number it turns out needed to bring about both arbitrary position and arbitrary orientation, in three dimensions, of a tool or part wielded by the robotic arm. Why, you may ask, does our BMW car need two more degrees of freedom than this minimum of six? The reason is that people are not rigid bodies. We bend and need support to sustain a pose defined by several, interconnected, and not-quite-rigid parts. Some robots actually have more than six degrees of freedom; many of them have fewer. The reason for the case of fewer is that general spatial motion is not always needed. Material-handling robots may never require poses that tilt a pallet, or box, say, off of an orientation that is parallel to the floor; such robots might have four degrees of freedom. They benefit from the cost saving of commensurately fewer servomechanisms and motors. Mechanisms with more than six, such as the human mechanism, not only can place a tool at an arbi-

Figure 2-1. Teach and repeat is part of your daily experience if you drive a car with seat memory. Rather than use the "teach pendant" common to most robotic arms, you probably used buttons and switches on the side of the seat to establish in memory the exact number of degrees of rotation of each of the seat's motors.

trary pose within their reach in fully three-dimensional space; we can also reach around things to do it. Canada's Special Purpose Dexterous Manipulator, designed for use in space, has more than six, for example.

How do all these servomechanisms work? On the face of it their job is daunting. Consider the case of a direct-current motor, the kind perhaps used for our automobile seats. Voltage applied to its armature, a winding of wires that uses Michael Faraday's famous principle to create torque, causes electrical current to flow through that armature. The current produces mechanical torque, which turns the motor shaft thus moving the seat up into position. And so the following might seem reasonable: just deliver the right amount of voltage to the motors, have a computer remember, say, just the right number of seconds, perhaps the exact number of seconds you used when you first taught the pose, and there you've got it – the seat rises just the right amount and on command. But it isn't so easy. You may have gained weight since you first taught the pose (after all you've stopped getting all that *manual* seat-adjustment exercise). Or maybe another driver is borrowing your car. Be she Twiggy or a pre-diet Anna Nicole the seat is supposed to recover exactly the angle you taught. And the voltage

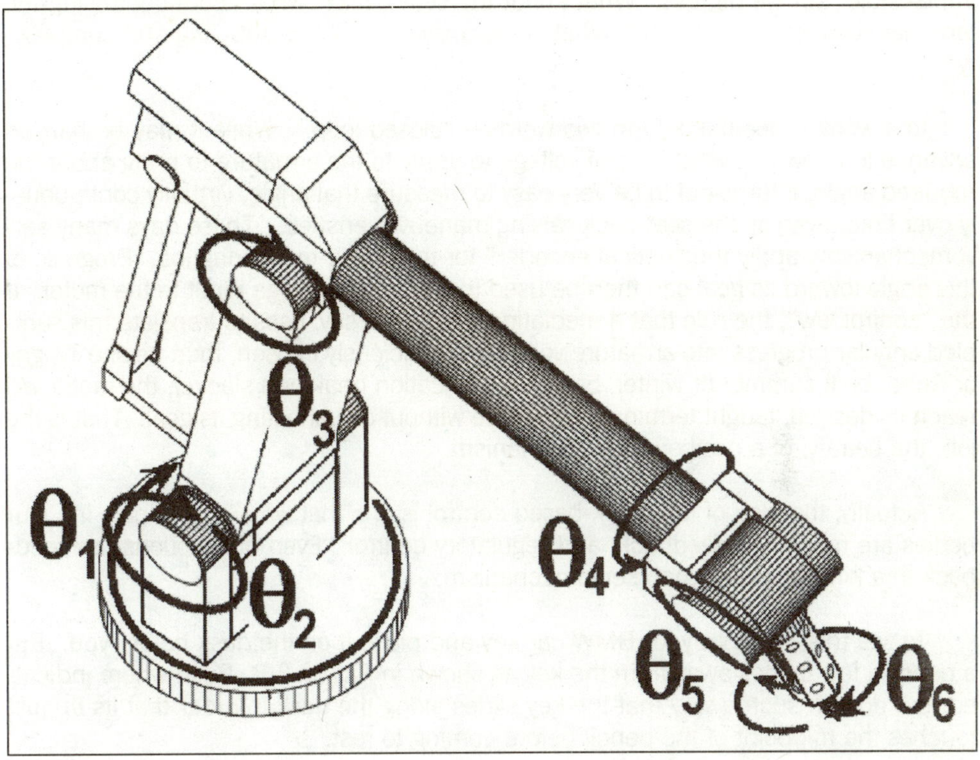

Figure 2-2. "Teaching" of this six degree of freedom robot is comprised of setting the angles, via teach pendant, which are needed for each stage of the task.

history over time needed to do this will differ according to the load borne. Worse still, all kinds of other, nearly imponderable factors influence the actual movement: That relationship between voltage and current? Very complicated (though clarified some by Farraday) when both are changing fast – and different depending on temperature inside the car. And when was the last time you lubricated that seat joint? That will have a *big* effect. In fact, all of the factors of load, inertia, current, friction, torque, and voltage "couple" – influence one another dynamically over time – in complicated ways whose effects accumulate during the motion to have a very big final influence on the terminal angular position of the seat back.

The very idea of what some call "open-loop control" – calculating in advance what actuator or motor input is needed to achieve especially position, position being the integral after all of velocity, i.e. more opportunity to accumulate unintended effects of any ignorance – becomes absurd. (That is why the technical offering of much of this book, advocating a *kind* of open-loop control for positioning robotic arms seems implausible; but we digress.) What about the seat back? What about controlling the terminal position of the motor? What is the answer to all this coupling and complexity?

In a word – "feedback." In two words – "closed loop." While it may be hard in advance to know just what kind of voltage to apply to the armature to bring about the required angle, it turns out to be very easy to measure that angle, virtually continuously over time, even as the seat-back-raising maneuver ensues. These days many servomechanisms apply the "optical encoder" for this angle-measuring job. Progress of this angle toward its goal can then be used to adjust the voltage input to the motor. If the "control law", the rule that a mediating computer, say, uses to translate this sampled angular progress into armature voltage, is judiciously chosen, then be she Twiggy or Anna, be it summer or winter, be you a lubrication freak or a slacker, the motor will reach its desired, taught terminus every time without overshooting its goal. That is the job, the beauty, of a position servomechanism.

Actually, the idea of feedback-based control is one that we all participate in. Our bodies are marvels of feedback, and regulatory control. Even our motions use feedback in a kind of closed-loop servomechanism.

To see this, take out your BMW car key and place it on the desk before you. Put a pencil a few inches away from the key as shown in *Figure 2-3*. Push where indicated, but do it in such a way that the key slides along the desk in order that its tip just touches the midpoint of the pencil before coming to rest.

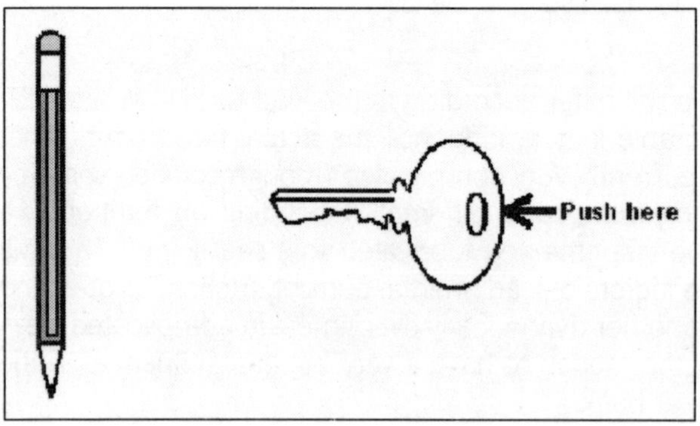

Figure 2-3. The remarkable ability of feedback, in this case visual feedback, can be seen with a simple experiment.

There you did it! No working out the mass of the key. No open-loop calculation. You just kept your eye on the thing as it moved. It wanted to rotate but you wouldn't let it. You shifted your finger action gradually to keep it on course. You managed three degrees of freedom of control: two of position (a component along the length of the pencil and a component perpendicular to this length) and one of orientation. And you did it all by feeding back information from light reflecting off of the key and into your eyes. Importantly, you did this in the reference frame of your own visual sense. No calibration needed here to learn where the target pencil was relative to any particular "world coordinate system" with origin, say, at the corner of your desk. If the key tip made it to the pencil midpoint in the reference frame of your personal visual sensor, then it made it, period. (The seductive appeal of visual servoing as a method to control robots beckons. But again we digress.)

The main point here is that the physical factors coupling together to influence the trajectory of the key are if anything more complex than those of the driver's seat. Dry friction, or Coulomb friction as it is called, dominates the motion you produced, and it is notoriously difficult to model or predict; it was this and myriad other vagaries of actuation and subtle physical interactions that your use of visual feedback overcame. You had some kind of control law at work, but this would have been different if someone else had done the same task. Yet both control laws would succeed. Maybe not on the same timetable, maybe not following exactly the same slightly crooked line on the desk, but both would get it there. The sine qua non is feedback and use of the "closed loop" – constantly correcting actuator input to ensure progress toward the goal.

So it is with the rotational servomechanism of the motor, the one that controls the backrest orientation of the driver's seat and the one that controls one of the six motors of the robot in *Figure 2-2.* Such servomechanisms entail one degree of freedom of control each. But put them together in a multiple-motor device such as a driver seat or a robot and you can reach terminal states that entail many specifications of position and/or orientation. A teach pendant can be used to specify all six components of position and orientation as may be required to perform complex and precise spot welds on automobile frames such as indicated in *Figure 2-4.* Efficacy of this approach, however, depends upon returning each new-vehicle frame to just the point in space occupied by the prototype vehicle, the one used when the angles were taught.

That is unless you are very fussy not just about where you terminate the maneuver, but also the details of the route taken to position the robot's tool in space. There is a difference between high precision at a terminal rest state, such as just before the emission of the spot weld, and precision in the sequence of poses that are passed through, during motion, leading up to that point. To appreciate this difference, it is worth looking at a "block diagram" of the position servomechanism itself as shown in *Figure 2-5.* This is a common representation of the principal elements that affect the dynamic response of the machine. For any well-designed "control loop" there is no issue of reaching the taught terminal pose, or for one motor/joint, reaching and coming to rest at the taught number of degrees of rotation; this is easily enough achieved.

Figure 2-4. The beautifully orchestrated coordination of mechanical spot welders in a factory is not in its essence so much like a well-disciplined, precision corps of human workers. It has much more in common with the movement of your memory driver seat as you turn the key. The factory's orchestrated movements are, for better or worse, the result of servomechanisms recovering previously "taught" angles of rotation. *Reprinted by permission from Robotics Online (www.roboticsonline.com), the official Web site of the Robotic Industries Association.*

But what happens en route to coming to rest can be very tricky to anticipate and, it was once thought, to control.

The top blocks or processes of *Figure 2-5* all entail differential relationships: It is the *rates of change* of values of interest, such as the angular position of the motor, θ, that respond to the various input quantities. Everything affects everything else. What is sure is that a well-designed control law will not allow the actuator or motor to quit until it has brought the terminal angle θ to its desired "reference" value θ_r.

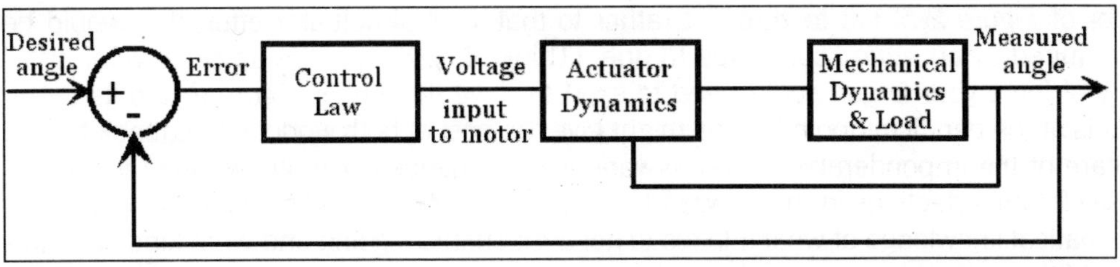

Figure 2-5. Diagram of servomechanism for one degree of freedom. A complicated set of relationships combine to produce the final, actual response of the measured angle. These relationships are mostly differential; and though they depend upon a variety of factors, the control law will not allow the actuator or motor to quit until the actual, terminal, measured angle exactly matches the desired angle. Usually, the control law is such that only the presence of nonzero error will get and keep the motor driving. An effort to exploit potentially knowable torque-response relationships thereby avoiding much of that error, even with the complexity of multiple servomechanisms influencing one another in a robot, prompted many in the 1980s to investigate the feeding forward of a computed-torque control.

En route to the terminus is another story however. If, for instance, there is a large initial gap or difference or error to overcome, the actuator will push hard. This in turn forces a fast but difficult to predict response. Part of the problem is the variable, physical load itself. As with the variable mass of the BMW car driver, so too a robot may be called upon to move loads of various magnitudes. Adding to the complexity of both driver seat and robot responses is the fact of simultaneous action of several motors. While not reflected in the single-servomechanism block diagram of *Figure 2-5*, it is not difficult to imagine that each of the various joints' movements of *Figure 2-2* will have an influence on the dynamic response (though, for a well-designed feedback-control system not the terminal value) of each of the six joint rotations. This makes the actual trajectory of a robot's end member through space difficult to anticipate.

This particular problem gave rise to a historical curiosity in the late 1980s and early 1990s. Researchers recognized that the coupling or dynamic influence of one joint's action on another could, from the point of view of an analyst, be predicted. If it could be predicted, then perhaps large error en route to a target could be precluded through a judiciously chosen, smart controller.

After all, why should one have to "wait" for an error that one could perfectly easily predict to arise before commanding the actuator or motor to deliver remedial torque control? Why not be proactive? Compute the torque ahead of time, based on Newton's laws and knowledge of the desired motion and simply "feed forward" this knowledge to the actuator rather than restricting actuation to the dictates of a usually behind "feed back". This was not advocacy of "open-loop control" mentioned earlier. The idea was to add the usual feedback-based control calculation (the input to the actuator of *Figure 2-5*) not to zero but rather to that level of actuator effort that would be required by an *apriori* computed torque. (Think of figuring out in advance what history of force your finger would need to exert to drive the key of *Figure 2-3* to its terminus at the pencil midpoint). This might give the best of both worlds: Feedback to take care of the imponderables such as various disturbances or unknown load variability, say, or imperfections in the forward-kinematics model; and judicious use of our mathematical knowledge of what it takes to get a mechanism going and what kinds of internal forces should, according to our best models, sustain its motion along a desired trajectory. It seems like a good idea now and it was irresistible to investigators in the 1980s. So predominant, in fact, were "computed torque" methods in those days that for many they became synonymous with robotics research.

There has been significant attenuation of such efforts and few if any actual robotic arms use the ability to compute torque histories required for a given motion. Why all the flurry of research activity with little or no consequence or follow-up?

Several reasons, probably: One is that conventional controllers do a good job - and they are getting better all the time. With many intermediate "via points" computed, usually with the robot's proprietary software, to interpolate motion from taught pose to taught pose, the robot's desired θ_r (see *Figure 2-5*) actually changes, smoothly, with time. (Note that most robots also allow the human teacher to select via points with the teach pendant. The robot controller will try to near these en route to the terminal rest position. This is very helpful for such trajectory-dependent jobs as spray-painting the complex contour of an automobile body for instance.) Control laws that use the sensed error at any moment to coax the joints to track in close conformity with the specified movement are well evolved and effective. Detracting too from the incentive to feed forward a computed torque is that some of the most influential forces during real-world motion are more like the Coulomb friction of the key experiment of Figure 2-3; they don't lend themselves to *apriori* modeling and analysis. Because real-world robots generally execute the same action maneuver after maneuver, the most effective way to anticipate joint-level torque requirements is to build in a corrective cycle based on previous maneuvers' actual error, rather than based on *apriori* kinetics analysis. And finally, near-perfect dynamical tracking precision is simply not such a pressing need in practice; it's not the main limiter of robot use. The reason robots aren't more commonly used where their physical abilities in fact are a good match to the task at hand has more to do with the limitation of teach-repeat itself.

When Twiggy gets behind the wheel and turns your key in the ignition, it's all well and good that all eight servomechanisms return to the seat pose that you taught. But she in fact likes to drive much closer to the steering wheel. Anna needs more clearance. The seat itself has all the versatility, mechanically speaking, to accommodate both of them. But exploiting this versatility using teach-repeat necessitates human involvement in reestablishing all the servomechanisms' memorized angles.

And so it is with the robot. Inherently, mechanically, it's possible to spray-paint a car that was driven into a room full of robots. But the actual motion of those robots can't be based upon simple repetition of a taught sequence of poses: The car's exact pose isn't known in advance, and the car's surface contours are variable. If one restricts the mode of robot control to teach-repeat, the *de facto* norm today, the applicability of general mechanical dexterity becomes reduced to a level which, in comparison with the vast range of prospects physically compatible with robot motion, is very nearly zero.

The alternative, of course, is to control the robot's dexterity in response to "as-located". This requires non-contact sensing – vision, in most cases. And it may require some human supervision, but not all of the detailed attention of traditional teach-repeat. Achieving this potential is the goal of this book. Many robot motions, as mentioned above, require far more than one taught point to be effective. The manipulated tool must move continuously in reference to an object's surface contour – as built and as located. This and other tasks are among those best achieved with the technologies herein.

CHAPTER 3

CALIBRATION AND THE 1980s:

BIG INSPIRATION, BIG PUSH, AND A SURPRISING IMPEDIMENT

The logic for robot use on an assembly line is compelling: human replication of the positioning of one object relative to another, workpiece after workpiece as the line moves on, requires mechanical dexterity and "hand-eye coordination". But endowing tireless, steady machines with guidance based on computer vision proved too difficult. In retrospect this failure has set the stage for robot use - or lack thereof - to this day.

Ask an engineer, and it's probably the answer you would get: How do you direct the internal rotations of a robot in order to cause it to add a part precisely onto an assembly of, say, an automobile? Likely answer: Use computer vision to locate the pertinent assembly juncture, and then use the geometric model (kinematics) to calculate the robot's internal joint angles needed to locate the grasped part at that juncture; then command the robot's joints' internal servomechanisms to move to the calculated pose.

It seemed straightforward. Faster computers - combined with the use of specialized "structured lighting" – had made it possible, in the controlled environment of a factory, to locate in 3 dimensions the *target* of manipulation, or "workpiece". Robotic arms with the needed dexterity, the ability to place an object into any position and orientation, had been around for a while. And competition from abroad had, by the 1980s, made itself felt in the auto industry. Costs, particularly labor costs, had to be contained; and quality had to be improved. Visually guided robots, it was thought, would provide much of the answer.

Not only had a workerless, "lights-out" factory been considered the next big thing by many auto-industry executives, the public too pervasively perceived an unstoppable march of robot use. Quoting from the 2004 SRI International final report, *Implications of Information Technology for Employment, Skills, and Wages: Findings from Sectoral and Case Study Research* prepared by Michael J. Handel, (http://www.sri.com/policy/csted/reports/sandt/it):

"The possible employment implications of industrial robots exert tremendous popular fascination. In 1986, a Roper poll found that 63 percent of adults thought that the use of robots on assembly lines would increase unemployment despite retraining efforts and 53 percent favored 'severely limiting' their use. A few years later (1989), a Gallup poll found that 52 percent of adults expected that robots would replace most assembly line workers by 2000."

But the twin problems associated with calibration - accurate and sustained camera calibration to locate the workpiece, and equally accurate robot-mechanism calibration to place the manipulated object there – made the expected transition to robots fail. The robots too often would simply miss. The decision of auto makers at the end of the 1980s to back off of the ambitious plan to automate to the extent of controlling their mechanisms using computer vision may, in retrospect, have been the nail in the coffin of widespread use of general mechanical dexterity and mobility.

Yet one must look hard to realize this. It's difficult to separate wishful thinking from fact sometimes. News stories, for example, concerning a new robot system or capability are today and were then frequent, leaving many misimpressions. Writing for *Advanced Manufacturing* in 2000, Todd Phillips thought he noticed that the robotics industry had turned a corner:

> "Every few years, a new technology explodes onto the scene, promises everything, dazzles briefly, and then fizzles. Robotics is one such technology that seemed to have it all, but then, for a host of reasons, never took off as quickly as expected and never caught hold across a broad range of industries."

For a mix of reasons, subtle reasons, reasons involving the public's predisposed interest, academe's and industry's vested interest, and that peculiar response evoked by anything that looks and seems to act human, there is a widespread deception or at least confusion.

Clues can be found from the experience of the automobile industry, the industry that far and away uses the greatest number of robots. The above-referenced April 2004 SRI report states:

> "General Motors (GM)'s experience in the late 1980s was one of the most widely publicized failures. GM responded to surging Japanese imports with massive investments in automated equipment (including industrial robots), believing automation to be the key to renewed competitiveness. However, the software and equipment were bug-ridden and failed to operate properly, and productivity remained low. In fact, horror stories of automatic equipment smashing into other equipment or work in progress and failing to deliver parts, paint sprays, or welds to the proper place were reported widely. By the early 1990s, even GM executives recognized the initiative was a failure. ('When GM's Robots Ran Amok,' *The Economist*, 8/10/91)."

After the dust settled, some new labor-saving mechanisms had come into use, but their function was to bear the weight of a load directed into place by the human worker's remarkable ability to guide closure visually.

Japan itself eventually reduced robot use. Writing for the *Associated Press* in October, 2003, Jonathan Fowler states: "Japan still remains the world's most robotized economy, home to about half the 770,000 robots working in factories around the world, the study said. But, with the Japanese economy continuing in the doldrums, the number of robots has dropped steadily from a peak of 413,000 in 1997, as companies choose not to replace some aging machines. Last year, the figure was around 350,000." Making the point, this drop in the use of industrial robots occurs even as the cost per robot unit declines and robot quality improves. In his 2000 *Advanced Manufacturing* article Todd Phillips writes: "Prices of robots are falling rapidly relative to labor costs. Robot prices in 1999 were 40 percent lower than in 1990."

Robotics' critical missing element can, with some reflection, also be noticed outside of the sphere of industrial production. Take that latest home robot you read about in this or that publication. The robot with arms and legs and, apparently - yes, eyes. Your first reaction when you saw the picture and headline was that that'd be great: A robot to pick up your room, to do your dishes, to take out the garbage. Then you read the news story more carefully: The machine is "friendly". That seemed good. "It waves." Okay. "It can move forward or to the side." Yeah, but what can it do that I *need*?

The trouble is that everything you want done requires visual coordination of robot extremities with objects in your world - or movement through and to such objects. Experts often assure us that it's really just a matter of cost; we need to be willing to pay more for effectual robots. But if it's a matter of cost, does the Sultan of Brunei have a robot that will pick up *his* socks?

What we really need is someone pointing out the Emperor's true attire. Perhaps it is here: Writing for *TechWeb.com* in 2005, editor-in-chief Fredric Paul discusses the recent RoboNexus International Conference & Exposition in Santa Clara, California.

> "Beyond the gee-whiz factor, somehow all this robotic brouhaha didn't seem to add up to all that much. Despite United Nations predictions of fast growth in robotics, [an expert from a technology firm] said high prices, lack of standards, power issues, and the need for more computational ability are holding back further acceptance.
>
> According to [this expert], though, the application of standard PC technology can eventually help move robotics past the hobbyist stage that PCs, themselves, escaped back in the 1980s. [The expert's company] is now pushing its standard PC boards into the robotics community, hoping that off-the-shelf components will lower costs and impose de facto standards.
>
> Maybe. In fact, we hope so. But, according to us, the problem is not about costs and standards. It's about what the robots do-or, more precisely, don't do. While the military robots may already offer a 'killer app,' the more commercial bots can, what-vacuum my house? Badly? Until someone comes up with a workable commercial and affordable robotic 'killer app,' I'm afraid these technoids are going to remain living in the days of the Apple II."

From the summer 2005 issue of the American Society of Engineering Educators, Noel Sharkey says: "Everybody wants to hear that robots are going to take over the world, but it's not going to happen. You get a lot of scientists, particularly American scientists, saying that robotics is about at the level of a rat at the moment. I would say it's not anywhere near even simple bacteria."

But what about those military-robot "killer apps"? Conveniently classified, but probably human-in-the-loop control?

Early in 2005 the federal government itself disabused an otherwise technologically informed group concerning robotics reality. Cries, it seems, for servicing the Hubble Telescope to extend its life were first met with government objections that the U.S. should not risk a second Space-Shuttle mission sending humans to Hubble. This, however, was followed by cries for nonhuman, robotic servicing. The realities of our robot-control limitations needed explanation. The Committee on the Assessment of Options for Extending the Life of the Hubble Space Telescope, National Research Council obliged, writing its *Appendix D, State of the Art in Robotics* to supplement the 160 page document *Assessment of Options for Extending the Life of the Hubble Space Telescope: Final Report.*

The authors of the present volume believe that the points in this document's *Appendix D* are important but should be commented upon. We therefore quote the report below interspersing our own commentary. Words from the report are in quotation marks and Italics. Our commentary is in regular type.

"Appendix D: State of the art in robotics.

Robotics is a field that has many exciting potential applications."

We do agree. One question to ask here is: Completely apart from the question of control, what is the ability of simply the hardware available today to achieve the range of dexterous/mobile ends currently achieved only with ongoing human involvement? Strictly from the point of view of the hardware, mechanisms easily built today could achieve more than half of what currently requires ongoing human involvement we believe. And due to advantages of steadiness, strength and speed that could be built in, in addition to the fact that the mechanisms need not be nearly so general-purpose as humans, there is a sphere of prospect for newly construed useful tasks that one day we think will dwarf even this subdomain - provided the control problem is solved.

"It is also a field in which expectations of the public often do not match current realities."

Perhaps an understatement.

"Truly incredible capabilities are being sought and demonstrated in research laboratories around the world."

The word "incredible" is a very interesting one in this context. Unlike other areas of technology our intuition for what is or is not credible is biased by the fact that robotics seeks to mimic or replicate human abilities, which of course we witness in ourselves and others every day. The ability of a human to clear a table in a restaurant is not incredible to most people. But while building a wheeled device with onboard arms mechanically capable of this task is now easy, the visual guidance required by the task is far removed from anything mainstream robotics could possibly achieve. Robots may in fact relatively easily do what seems incredible in a lab, or even on stage. The entertainment robots of Epcot Center fifteen years ago were, in this sense, truly incredible. The most gifted human entertainer could not approach what those robots do. But useful? No.

"However, achieving these capabilities with real robots in real environments faces many hurdles. It is true that robotic systems can be stronger and faster than humans, can go places too dangerous for a human to venture, and can operate without fatigue while performing highly repetitive and precise tasks. However, it is very difficult to build a mechanical device (e.g., a robotic arm) that has dexterity comparable to a human's limbs."

In most cases the mechanisms are not the impediment. Muscles and tendons may have advantages for a few things, but gears and motors are likely to have the bigger advantage for narrow applications. Look at cars vs. horses.

"It is even more difficult to build a computer system that can perceive its environment, reason about the environment and the task at hand, and control a robotic arm with anything remotely approaching the capabilities of a human being."

Yes, this is the problem. But even here, more detail would be revealing. The rationale for 1980s attempts to build a workerless factory was that most of the time there is no need for human workers to perceive very much in a general sense about their environment, or to reason about the task at hand. The only extent to which adaptability was needed was to adjust the joint motion of the dexterous robotic arm to position its tool, or part to be added to an assembly, by that (generally) small increment needed to accommodate as-located workpieces – as opposed to precisely prepositioned workpieces for which the teach-repeat action of today's robots works well. This could not be done reliably due to the vagaries of calibration of mechanism and optics. The above statement may imply that the cognitive abilities of humans that aren't replicated artificially are those that entail high-level reasoning and so are understandably difficult. The most profitable applications of robots, however, and this would include semi-autonomous servicing of the Hubble, will only require competence in the narrow sense mentioned above.

"Hollywood's depiction of robots often endows them with human-like intelligence and decisionmaking capabilities, but real robots fall far short of this image. A robot is simply a machine that 'synthesizes some aspect of the human function.'(J.J. Craig, Introduction to Robotics, Addison-Wesley, 1999)"

If you stop to think about it, how much does this definition really convey? Is a remote-controlled car a robot? A voice synthesizer? How about a drone spy aircraft? Or a pair of stilts? The definition doesn't really help answer. This is a symptom of much of what is wrong in the arena in question. There is human language and (as with the above) allusion to human experience on the one hand, and there is (as in any field) technical jargon on the other. One very large effort supported by the European Union, for example, noted that the robotics community had already managed to achieve "the intelligence of an insect." It went on to state the mission of its current project as moving this benchmark up several notches to realize "the intelligence of a preschooler." What does such language really convey?

"In general a robot involves some level of ***automation****, which is the attribute of being able to perform a task or a sequence of tasks and adapt to a well defined and predetermined class of variations. A robot may also exhibit* ***autonomy****, which is the ability to make decisions the way a human being might make decisions. However, the level of autonomy that has been achieved in today's robotic systems is no match for even the simplest decision-making capabilities of a human."*

This is certainly true. It does mask, however, the gap between robotics reality and the key ability to respond to small variability in target-workpiece position/orientation. Much of the profit of prospective robot use would be realized, including Hubble servicing, if we achieved just this.

"Many robots are teleoperated. In ***teleoperation,*** *a human operator controls the robot directly while monitoring some or all the information that the robot sensors acquire. Teleoperated robots have been used effectively by human operators to augment their skills or to be able to operate in remote, usually hazardous or inaccessible, environments. For example, the manipulators used on the International Space Station (ISS) and the shuttle are teleoperated. Surgical robots that allow surgeons to perform procedures while operating through tiny ports are also teleoperated. The key feature of teleoperation is that it exploits the perceptual capabilities and reasoning power of the human operator rather than relying only on the sensors and computers available to the robot. A key requirement for successful teleoperation is that the communication link between the human operator and the robot is sufficient to provide enough information for the remote operator to make decisions and to issue appropriate control commands in a correct and timely manner. Teleoperated robots typically require and exhibit very little autonomy because of the presence of the human operator in the loop."*

The main reason people and organizations decide first and exclusively to use teleoperation as the mode of control is also the biggest impediment to its use. It gets back to the simple ability to bring about a geometrically desirable quality of closure between end effector (or object in the end effector's grasp) and workpiece. It isn't primarily a matter of lots of high level abstract decision making, the kind only done by an expert human (expert systems are an actual success of artificial intelligence), but rather this most mundane, seemingly trivial of abilities. This ability is what makes for a skilled crane or backhoe operator. It is the ability that in teleoperation is reduced by the time delays and indirect visual access. CSM with its precise autonomous control at this level is, we believe, the answer.

"It is useful to look at some well-known applications of robotics to understand the difference between automated, autonomous, and teleoperated robots. One of the most visible and successful application of robotics is in factories and on the shop floor. Here, reprogrammable, multilink robotic arms have replaced special-purpose machines to perform precise and quick repetitive operations, such as pick and place tasks, for handling parts and tools and for assembling parts. The advantage of using robots in these applications is that their reconfigurability and flexibility make it possible for one assembly line to be multifunctional and to be adapted for a range of parts or products. However, a production facility or a factory is typically a highly structured environment. Precisely manufactured parts arrive on schedule at predetermined positions and orientations for robotic operation, and all operations are, for the most part, predictable. Once a robot is programmed, very little "intelligence" or autonomy is required of the robot for it to perform its limited set of functions. Very little adaptation to uncertainties is required. In spirit, these robots are closer to machines like programmable looms or dishwashers than to Hollywood's R2D2."

This recognition is the one thing that, if it were appreciated widely, would be the single biggest education for the public – particularly if it was also disclosed that teach-repeat represents the dominant mode of use of what are referred to as robots.

"Another recent, very visible application of robotics is the pair of Mars Exploratory Rovers (MERs), Spirit and Opportunity. These very successful mobile robots exhibit multiple levels of autonomous or semi-autonomous operation. These rovers have sensors that provide information about the environment in which they are operating, about their position in that environment, and about the status of the task they are performing. The sensors provide information to computers, which reason about the state of the robot and the environment and calculate the commands sent to the robot's actuators to control its motion and activities. Some of this reasoning is done onboard the vehicle. However, much of the high-level reasoning and decision making is done by the remote human users, albeit infrequently because of the time delays associated with communication between the rovers and mission control on Earth. For example, remote human users set the science objectives (e.g., on which rock to place an

instrument) and issue high-level commands (e.g., "go to that rock"). The rovers then execute these commands using onboard sensors and computers to determine and follow safe paths through the terrain. Importantly, the onboard autonomy is limited primarily to the specific tasks of navigation and instrumentation placement. The rovers have some limited ability to adapt to operating conditions and the environment. When unexpected situations or failures are encountered, the rovers can stop and wait for the remote human users to issue a new set of commands. Human users can also make the decision to send new software to the rovers or patch software bugs that may be discovered during the mission. Thus, while these robots are not, strictly speaking, teleoperated, there is an element of teleoperation in the functioning of these rovers. At the same time, the rovers exhibit a significantly greater degree of autonomy than the automated factory robots discussed earlier. This combination of autonomy with an element of teleoperation is often called supervised autonomy."

This is a good example as there was one task in particular that required especially precise dexterous positioning: locating a camera for taking close-up images of natural Martian surfaces. The close-up camera's depth of field is just 3mm. The precision required was tight in the direction of the focal axis; thus the distance separating the rock from the positioned body had to be tightly controlled. The autonomous sequence of events involves additional hardware and preliminary motion, and to that degree makes the point: A touch probe was devised that reached out slowly until contact was felt. Only then was the small incremental distance needed for further extension of the arm calculated and executed. This procedure, however, offers almost no counterpart for tasks that require closure with more degrees of rigid-body position/orientation prescribed. These latter capabilities will be required we believe by future remote tasks. Moreover, after 180 sols the calibration of arm and optics that had been established drifted to the extent that an unanticipated corrective measure was adopted: end-member points were located in the camera images back on earth to re-establish calibration. This repeatedly used approach – similar to CSM in that the source of drift, optical or mechanical, is not isolated – permitted a return to remote-site hand-eye accuracy within the 1cm requirement. CSM, we claim, would allow for the still tighter subpixel level of precision in all components of relative position without non-optical sampling, and its robustness to drift is automatically ensured in real time at the remote site; there is no need to make corrections based upon images returned to earth.

"There are many remotely operated vehicles like Spirit and Opportunity that have been deployed on Earth. Rovers have been used for nuclear reactor inspection at Three Mile Island and have been deployed by the military for de-mining in Bosnia and for reconnaissance in caves in Afghanistan. In Iraq teleoperated rovers with manipulators are used for disruption and disposal of improvised explosive devices. Robotic submersibles have been used in the deep sea for exploration tasks by the marine science community, for inspection and maintenance tasks by the oil industry, and for sal-

vage of wrecks like the Titanic. The level of autonomy employed in these devices varies. It is not feasible to teleoperate the MERs because of the time delays associated with communications; hence supervised autonomy is used. It is feasible, however, to teleoperate a vehicle driving over a minefield. Thus a military robot clearing mines through a minefield may not require the level of autonomy that the MERs require."

Human in the loop control, however, is very limiting. It is autonomy at the joint-control level that is needed and is a big part of the message of the present volume.

"Robots can also be seen in the service industry. There are commercial products for vacuum cleaning, for mowing lawns, and for assisting people with disabilities. Humanoid robots are being developed for entertainment. There are many sophisticated toys that employ robotics technology. Amusement parks use programmable, articulated mechanical devices to mimic biological motion. While many of these applications provide successful examples of autonomous operation, there are no examples of dexterous manipulation."

Reliance on a random movement in certain respects serves as evidence of the paucity of more useful solutions.

"Deciding what tasks can or should be performed autonomously by a robotic system depends heavily on the details of the specific mission. Further, enabling those autonomous operations requires an extensive, dedicated research and development program, which begins in the laboratory and culminates in field demonstrations before actual deployment on a mission. For the MER rovers, autonomous navigation was identified as having significant mission benefits and was achieved only after years of focused research and development, such as identifying obstacles using computer vision and relative state estimation using wheel, inertial, and optical sensors. Manipulation with robotic arms is a very different type of task and requires a similar, focused development activity if it is to be automated at any level. Robotic arms have been used extensively on the shuttle and on the ISS to perform assembly-class operations, but up to now all of these operations have been done in a teleoperated mode with no autonomy. (Astronauts at the site monitor and control the motions of the arms directly. The information they use includes direct visual observations plus views from video cameras and readings from joint angle sensors mounted on the arms. They control the motion of the arms using a joystick to issue commands that control the torque produced by the motors embedded in the arm.)"

Understanding this would be the second most useful way to educate the unsuspecting public on robotics.

"Significant training of the astronauts is required to qualify them to use these robotic arms."

And there are profound limits even with the significant training. Hence human "extra-vehicular activity" (EVA) is required.

"Automated rendezvous, capture, and grappling of HST and robotic servicing with dexterous manipulators cannot be performed via direct teleoperation because of the time delays in the communication link between the orbiting robot and the ground station. (The delay expected between the ground and HST is approximately 2.5 seconds. In order for teleoperation to work successfully, the information supplied to the user must be sufficient and timely. When controlling a dynamic system, excessive delays in the information transfer between the device and the user can cause the system to go unstable. In particular, the can be achieved by limiting the speed of robot motions during teleoperation. However, if force feedback is used, even delays of a fraction of a second are known to cause instabilities during teleoperation and pose difficulties for a human operator. Force feedback is needed for inserting instruments into the HST, and for mating and de-mating of connectors.)"

An interesting historical note. In the early 1990s one of this book authors had discussions with several engineers who had first-hand knowledge of the coming to an end of the automotive industry's plans for a workerless factory. They indicated that it was the inability of the calibrated-vision-guided robotic placing of a wheel and tire assembly onto a brake plate. It seems that, as indicated, people do this using a good deal of feeling around. Force control lets us back off if there is resistance and rotate or otherwise jog the wheel until it slips into place. The robots wouldn't do this.

This was taken as the initial task for CSM and several videos of it can be seen on the University of Notre Dame Web site http://www.nd.edu/NDInfo/Research/sskaar/Home.html. The point is that the artificial solution shown there involves a kind of pure motion that just goes straight in. Deviation from a perfect trajectory is so small that the mere passive compliance of the arm is enough to guide the assembly into position. So often the perceived need for force control, especially with the assembly or unity of two objects into a kind of latched connection, is only a concession to imperfect physical positioning in the first place. The teleoperation therefore may be thought to be inadequate due to problems with time delay and force control, or it may be inadequate because the distant human operator, no matter how slowly the maneuver unfolds, cannot deliver high precision in six degrees of freedom.

"Supervised autonomy is the appropriate mode of operation for the robotic servicing mission. It allows shared control where the onboard computers can control the motion of the arms and effectors based on sensory information while human operators on the ground can make mission-critical decisions. However, the successful implementation of supervised autonomy requires that the manipulators, sensors, and control software be sufficiently sophisticated to perform assembly and disassembly tasks in an environment that is not well structured, unlike the structured environment

of the factory and the shop floor, for example. It is also important to note that although supervised autonomy has been extensively studied in research laboratories, its robustness and reliability for a mission as complex as the HST servicing have not yet been verified. There are very few examples of field-tested space operations involving manipulation or assembly with autonomy or supervised autonomy. In 1970, rendezvous and capture with a noncooperative target were performed by the Soviets with a human operator in control and without any communication time delays. In 1998, collaboration between ESA and NASDA produced a moderately successful demonstration on the Japanese Engineering Test Satellite (ETS) VII. This involved manipulation of a 2-meter-long, six-degree-of-freedom manipulator arm attached to a 2500 kg satellite with the coordinated control of the manipulator and the base. The ETS VII mission demonstrated autonomous rendezvous and capture of a target satellite. However, in this demonstration, the target was specially designed for capture, with appropriate fiduciaries for relative orientation, positioning, and capture. Thus the proposed HST robotic servicing mission will require the development, testing, and validation of new software and hardware, which would advance the state of the art of robotics technology."

We strongly feel that human supervisory control needs a combination of attributes afforded by the present technology. These include robust high-DOF relative positioning precision of two bodies – precision that benefits from samples en route to refine estimates of each participant camera's "camera space kinematics". It also requires an intuitive yet versatile means by which the human can convey the requirements of the maneuver to the remote system. For this the advantages of point and click surface point selection followed by remote, automatic convergence of a laser spot onto the selected surface junctures (a highly robust process) is uniquely suited.

What, though, about the academy? If industry, domestic/service robotics, and government come up short, surely the academic world, with its tens of thousands of scholarly papers and other works over the course of the past twenty five years, has answers regarding the proper direction for the future. Two broad strains of main-stream-robotics research that in the authors' judgment really attempt to address the serious impediments to widespread use are discussed in the next chapter.

CHAPTER 4

THE 1990s:
VISUAL SERVOING AND BEHAVIOR-BASED ROBOTICS

These two concepts, each in its own way, appeal to our sense of how natural systems work. Although the term "visual servoing" evokes ideas of the servomechanism, a stalwart of man-made control (see Chapter 2), the domain of visual servoing's closed-loop event is the reference frame of the visual sensor rather than any absclute physical reference frame - clearly a modus operandi of nature. Behavior-based robotics harkens back to nature's pragmatic way of realizing objectives without masterminding an immediate environment before action begins. Sometimes a well-selected set of programmed-in rules - possibly evolved from trial and error - as to how actuators should respond to specific sensory input is enough to realize useful and effective overall behavior. Why did these good ideas produce so little fruit despite great effort?

Two very different objects of robot-control research became particularly important in the nineteen nineties. Each had as its objective bringing about a degree of robot usefulness not seen outside of the limited, antiquated, but always reliable mode of teach-repeat.

Visual Servoing

Consider the needle-threading task of *Figure 4-1*. Putting aside temporarily the three-dimensional character of this real-world task, it is easy to see that the person is making increments in her own joint angles or degrees of freedom in response to the perceived gap or separation between thread tip and needle eye. Closure of this gap in the visual reference frame is her mind's objective. A very similar idea lies behind one form of visual servoing, a form where cameras are stationary and movement of the robot is directed to bring about closure between objects to be joined in these stationary reference frames. A second form, involving what's called the "eye-on-hand" configuration, might better be imagined by thinking of "bobbing for apples" with one's eyes open. Closure between mouth and apple in the reference frame of the visual sensor is the objective, but in this case the visual sensor or eye moves jointly with the engaging mechanism or mouth.

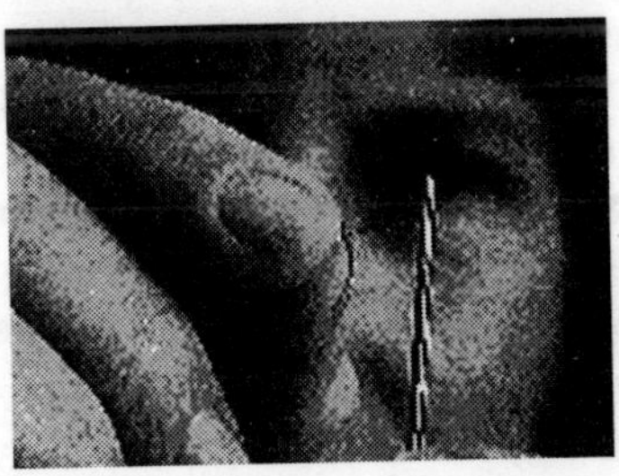

Figure 4-1. When threading a needle we don't think about the absolute location of needle *or* thread – just where each is in relationship to the other, *in our own visual frame of reference.*

Concentrating for the moment on the needle-threading case - eye *off* hand: how did engineers and researchers of the nineties bring about closure of two objects, one of them in the grasp of the robot, using the ideas of the "servomechanism" as discussed in Chapter 2? The actual working out of the rules of the closed loop - the "control law" - of visual servoing took any one of a number of possible forms, each with its own stability and performance characteristics. But all of these forms (and this part would be true of eye on hand as well) used what was most often referred to as a "Jacobian". Named for the French mathematician Carl Jacobi, this is really a matrix, an array, of partial sensitivities - approximations to what is called partial derivatives - which are needed to ascertain the partial contribution a slight movement in any one of the robot's joint angles would make in closing the gap *in the image* between objects to be joined.

The idea is that the image-plane or camera-space gaps that we are trying to drive to zero are responsive to incremental or partial motion of each of the robot's joints. If we could determine what these small responses are, at least approximately, then we could move the robot's joints accordingly, making progress in the effort to drive the camera-space gap or "error" to zero. And in the context of continuous feedback and closed-loop control, the sine qua non of any servomechanism, progress is what counts: Zeno notwithstanding, if with finite speed we continue to reduce the error, we will, in a finite amount of time, get where we're going. In *Figure 4-1* that means closing of the gap to zero between thread tip and needle eye – in camera space.

A good illustration with uses all its own - though not quite real visual servoing

Particularly in view of the complications of three dimensions, it is difficult to illustrate visual servoing in a way that could be easily replicated experimentally by a reader. However, there is a particular problem that has all the key attributes of visual servoing - one that turns out to be not only easily achieved by a novice but also of importance to one of the new-DNA technologies presented herein: point and click camera-space manipulation (Chapters 6, 8). This application entails determination of a Jacobian that relates increments in mechanism joint angles to increments in the movement of the manipulated "body" in camera-space. In this case, making life relatively easy, the body is a point, the center of a spot made by a small laser pointer. Think of a spotlight man trying to zero in a very small spot of light onto a tiny on-stage prop; or a lecturer trying to highlight a small subscript on his slide presentation using the laser pointer in his hand. Consider the system of *Figure 4-2.* As noted later, particularly in Chapter 8, the authors suggest using human visual judgment and a point and click form of graphical user interface (GUI) to select target surface-point position in camera space (the $\mathbf{x_c}$,$\mathbf{y_c}$ domain of *Figure 4-2*). Such a selected target surface point is represented by the white X inside the white circle in the figure. With this target in hand, what kind of information is required, and what kind of control strategy could be used in order to move the two selected degrees of freedom θ_1 and θ_2 of the robot in a direction that is consistent with camera-space movement of the laser spot toward its target?

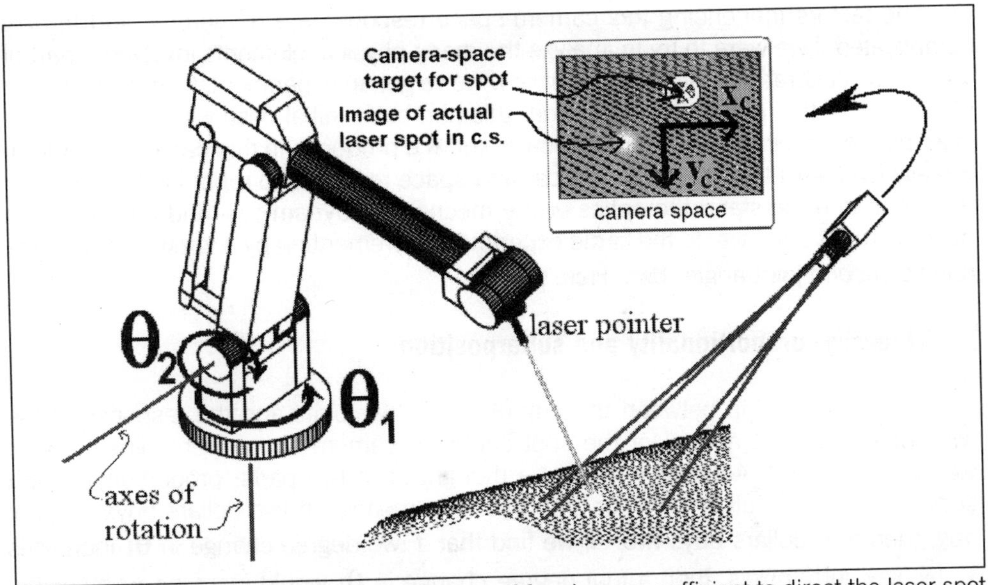

Figure 4-2. Only two of the robot's six degrees of freedom are sufficient to direct the laser spot onto the desired physical surface point. Shown at work here are the first and second joint rotations although almost any pair would suffice provided their axes of rotation are not nearly parallel. The objective is to drive the image of the laser spot to the point where it collocates with the target juncture *in camera space*, as shown.

Consider a small experiment. Suppose that, from the robot pose shown in *Figure 4-2*, the robot's #1 servomechanism (see Chapter 2) is commanded to increase the angle θ_1 by a specific small amount, let's say two degrees. Prior to making this movement, we use special software, as described below and in Appendix D, to register the coordinates of the laser-spot center in camera space; that is, we determined $\mathbf{x_{c1}}$, $\mathbf{y_{c1}}$ for the first laser spot as indicated in *Figure 4-3*. Similarly, subsequent to the motion of joint 1, we made a sample of the spot's center in its new camera-space location $\mathbf{x_{c2}}$, $\mathbf{y_{c2}}$. In accordance with *Figure 4-3*, we then observe how much this spot moved in response to the 2-degree movement of joint 1 of the robot. Response in the camera's $\mathbf{x_c}$ direction is given by $\mathbf{x_{c2} - x_{c1}}$, while response in the $\mathbf{y_c}$ direction is $\mathbf{y_{c2} - y_{c1}}$.

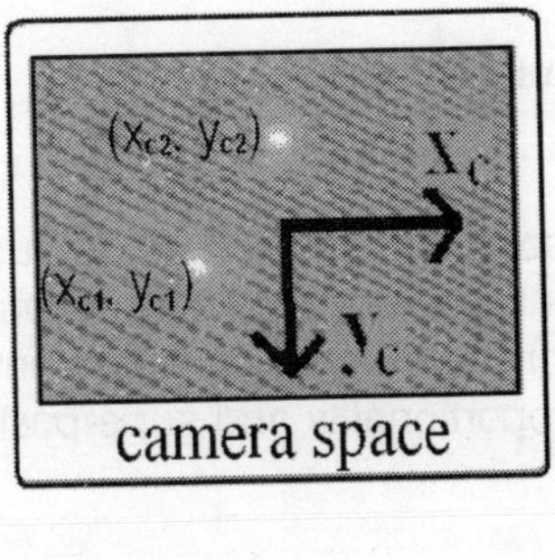

Figure 4-3. The laser spot is detected in camera-space at the indicated "1" position when the robot is in its base pose of *Figure 4-2.* After a two-degree movement of joint 1, the camera-space location moves to the "2" position shown.

The factors influencing this camera-space response are numerous, and terribly complicated if we were to try to analyze the many physical elements involved – part of the reason calibration is so tough. But now we've got an experimental idea of just what movement of this particular angle, from this particular initial pose with this particular laser pointer reflecting off of this particular surface produces in this particular camera. It really isn't hard to arrive at such a camera-space response to robot-joint action – no real need to understand the optics or the mechanism dynamics – and it turns out to be very useful once we do the same experiment incrementing by a known amount the robot's *second* joint angle, θ_2. Here's how.

Linearity: proportionality and superposition

If the relationship between movement of our two joints and the response of the two components x_c and y_c of the spot center in camera space were "linear" what would that mean? Actually, it could be thought of in two parts: proportionality and superposition. We all know what proportionality means: If two dollars buys one hot dog, then four dollars buys two. If we find that a two-degree change in θ_1 increases x_c by, say, forty pixels, then a four-degree change in θ_1 would increase x_c by eighty pixels. That's proportionality, nothing too new. But what about superposition?

Suppose you knew that for every marathon you ran you: 1. increased your life-span by one month, and 2. decreased your weight by five pounds. Suppose further that you knew that for every entire cheesecake you ate you: a. decreased your life-span by one fourth of a month and b. increased your weight by two pounds. Now your goal is to increase your lifespan by six months and increase your weight by three pounds. How many cheesecakes should you eat, and how many marathons should you run?

If these phenomena are linear, and superposition holds, a simple mathematical formulation can produce the desired answer. Let's let m be the required number of marathons and n the required number of cheesecakes. Then:

$$\begin{aligned} &\textbf{Six months} = [\textbf{1 month /1 marathon}]\ \textbf{m} + [\textbf{-.25 month /1 cheesecake}]\ \textbf{n} \\ &\textbf{Three pounds} = [\textbf{-5 pounds/1 marathon}]\ \textbf{m} + [\textbf{2 pounds/1 cheesecake}]\ \textbf{n} \end{aligned} \qquad (4.1)$$

By solving these two equations simultaneously it looks like you'll want to run m=17 marathons and eat n=44 cheesecakes. If you think that everything else in your life is unaffected by the number of cheesecakes you eat or the number of marathons you run and you trust the linearity of the relationships (proportionality and superposition), and if you think that the experimental results that went into the bracketed terms are accurate, you could just get it over with in one "open-loop effort" (no measurements, no feedback) – run the 17 and eat the 44, and that would be that.

An intermediate step toward servoing; how feedback can compensate for imperfect knowledge

But you're a little skeptical. You doubt that the experiments that went into all those individual ratios inside the brackets (they would be the four elements of a Jacobian, by the way) are all that wonderful. And you're suspicious that these phenomena may not be linear - that neither proportionality nor superposition quite holds. Here's where the beauty of feedback comes in: Suppose that you have the ability to monitor your progress toward your weight and longevity goals. After eating, say, half - **n/2** - of the cheesecakes and running **m/2** of the marathons, not only can you weigh yourself to see how things are going in that department, you can also buy Dr. Anne U. Itee's do-it-yourself longevity kit to see how many months this initial effort has extended your lifespan.

The Jacobian's forecasts are, as you'd suspected, quite a bit off. Rather than extending your lifespan halfway toward your six-additional-month goal you have actually extended it a full four months, a month more than you would have expected given those Jacobian elements, linearity and such. This leaves you with just two months more of longevity extension to go, not three. It's a good thing you stopped to use feedback rather than doing everything at once. In the weight department the predictions have done a bit better: you're up one and a quarter pounds, close to the goal-midpoint of one and a half.

Now you start to wonder. You're almost sure you heard Dr. Anne say somewhere that the cheesecake effect on longevity varies a lot from individual to individual. Maybe, before taking the next step in your program, you could use your results from the first half of the program to improve upon, personalize, or make more "locally" valid, the four elements of the Jacobian. You know that just one personal experiment taken by itself - since it provides just two new equations, one for weight and one for longevity - isn't enough to solve for all four individual numbers of the Jacobian. But shouldn't this new, most relevant-of-all information be able to help? Maybe to modify the original Jacobian more toward the numbers that are right for *you*? And couldn't we somehow use the fact that Dr. Anne says the most suspect of the four original numbers is the one that relates cheesecake to longevity? (Why weren't *all* doctors like this woman?)

In fact there is a systematic, "optimal" way to achieve just these Jacobian-improvement objectives, using just the kind of local information provided by your first test, together with Dr. Anne's insight regarding which of the four elements of the

Jacobian may benefit most from corrections stemming from the application of new, if partial, data. They are discussed in Appendices A and D. But even though, if we were to use estimation methods, convergence to the desired goals would be accomplished more efficiently, the point here is that if we just stick with the highly imperfect, demographic-average-based sort of Jacobian that we started with, a judicious plan forward is still likely to converge onto our desired result, with great accuracy and "robustness" - or insensitivity to those assumptions that you've gotten wrong. That's the great thing about feedback.

There are lots of effective ways to use feedback; here's a simple one: Now that we have two more months of extra longevity to acquire and 1.75 pounds to gain we solve for a new m and n by modifying the left sides of *Eqs 4.1*:

Two months = [1 month /1 marathon] m + [-.25 month /1 cheesecake] n

1.75 pounds = [-5 pounds/1 marathon] m + [2 pounds/1 cheesecake] n

We solve for **m** and **n**, again take *half* these numbers, and then run and eat accordingly. We keep doing this until we get arbitrarily close to our goal. And provided this reused Jacobian isn't *too* bad, not only can we get very close, but we will not overshoot our goal. (Negative marathons would be okay, if we overshot, but having to eat negative cheesecakes would be cruel.) In a concession to Zeno, it's true that, done this way, it would take infinitely long, due to an infinite number of finite-time corrections, to actually reach the goal. But in the real world engineers would, after a few partial corrections, forego the halving of the n and m and take execution all the way in one final move. As this would be a small move overshoot would be negligible.

The same strategy applies to the camera-space laser-spot positioning task of *Figure 4-2*. For this case the counterparts to Eqs.4.1 would be

$$\Delta x_c = J_{11}\Delta\theta_1 + J_{12}\Delta\theta_2$$

(4.2)

$$\Delta y_c = J_{21}\Delta\theta_1 + J_{22}\Delta\theta_2$$

where $\Delta\theta_1$ and $\Delta\theta_2$ are the solved-for changes that, according to the Jacobian, would be required to close the gap entirely in camera space. J_{11}, J_{12}, J_{21}, and J_{22} are the elements of the Jacobian determined from the aforementioned tests involving small angular departure first of θ_1 and then of θ_2 from the base position of *Figure 4-2*. (*Appendix D*) Consider for example J_{11}: J_{11} is the response in camera space of the $\mathbf{x_c}$ component of the spot center, probably in pixels, to a unit change in θ_1, let's say measured in degrees. If, as suggested above, the test change in θ_1 is two degrees and the corresponding detected change in the spot center's x_c is forty pixels then J_{11} = 40 pixels/2 degrees = 20 pixels/degree.

The left sides of Eqs. 4.2 are the targeted changes in camera space for the laser spot center. *Figure 4-4* illustrates Δx_c in particular; it is the difference in the camera-space x_c direction between where you want the spot to go and where it is currently. A good approach for actually applying feedback in a finite number of discrete mechanism moves to achieve convergence of spot with target in camera space is to use the step-halving strategy of the runner/eater above: With simultaneous-equation solving of Eqs. 4.2 for $\Delta\theta_1$ and $\Delta\theta_2$, command the two robot axes to advance in just *half* these computed increments.

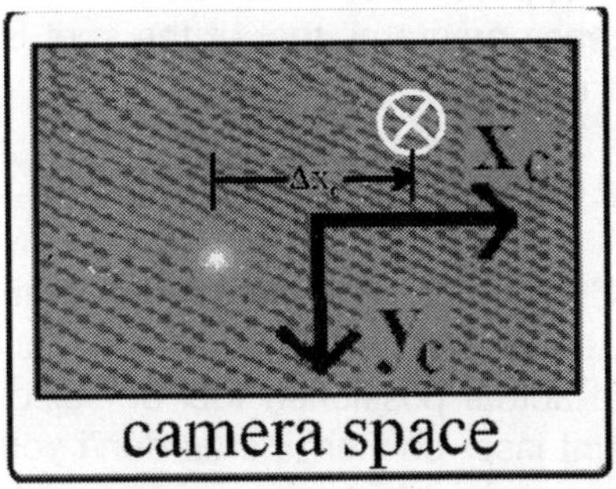

Figure 4-4. The component in the camera-space x direction of the needed correction for laser-spot-center positioning. A similar component exists in the camera-y direction. Both of these can be found using: **1.** Knowledge of the human-selected target location for the spot, and **2.** Automatic image analysis as per *Appendix D* to find the two camera-space components of the actual, current spot location.

Because only two degrees of robot freedom are needed here, a reasonably priced pan/tilt unit is suggested rather than a higher-degree-of-freedom robot for directing the laser pointer of *Figure 4-2*. As mentioned above, the resulting GUI/spot convergence subsystem will be a very useful tool described later in this volume in realizing the vast prospect of human supervisory command of autonomous robot execution. Using the camera-space-objective-realization method of camera-space manipulation, which is contrasted with visual servoing in Chapter 6, results in robust and precise robot control for achieving the human-designated objectives. This combination is the central device by which the advantages of autonomous artificial dexterity are brought to myriad new applications. We believe it will be the principal enabler of Alvin Toffler's "third wave" partly because, as discussed in Chapter 8, it is a form of human supervision to realize precise mechanism performance that is completely compatible with the internet as it is today.

Using the strategy of image-differencing together with an image template as per the discussion of Appendix D, the image-location components of the current spot center can be found from grayscale ("black and white") camera data. Various devices can provide suitable output including digital cameras or the combination of an analog video camera together with a frame grabber. As of this writing the latter is the more cost-effective way to go.

One note of caution as you try this exercise of converging a laser-spot center in camera space onto a preselected camera-space juncture: While it is true that convergence will usually occur even if the Jacobian is quite far from local reality (i.e. the actual, operational input/output ratios), the prospect of the laser spot dropping off one surface and falling onto a second some distance behind or ahead of the first surface upon which it fell means possible abrupt changes in these true i/o ratios. Such abrupt changes brought about by surface discontinuity can slow convergence. Also, particularly if the camera of interest and the laser pointer itself are separated physically, as is the case pictured in *Figure 4-2*, there is a second danger: The laser spot, especially again if it drops off one surface to fall onto a second, significantly separated one, can become hidden from the camera's view. The likelihood of this can be reduced by locating the camera in whose frame convergence is sought physically near the laser pointer. As a matter of pragmatic engineering, too, it is possible to build into the convergence algorithm "bells and whistles" — contingency programs that, if the spot is not detected, cause the laser pointer to move about until it is.

The Achilles heel of visual servoing

Actual visual servoing, however, usually entails continuous movement rather than the stop-measure-calculate-start motion discussed above. Motion continuity is particularly important when a heavy robot is guiding the actual positioned tool or object rather than a small pan/tilt unit directing a lightweight laser pointer (inertia don't you know). Such continuity would require continuous or nearly continuous (i.e. frequent, if discrete-time) feedback of the camera-space errors.

To create the counterpart to visual servoing for continuously gaining longevity and weight using the marathon/cheesecake method, you would need to use Dr. Anne's latest device (so creative!), the "real-time actuarial monitor". Just hook it up to one ear lobe and your current activities' expanding or contracting effect on your personal longevity can be viewed on a small wrist monitor. Just watching those years of future life grow is a powerful motivation to keep running. Fortunately the same monitor can be hooked up to a small sensor in your running shoes such that your current weight can also be read. None of this interferes either with running or grabbing more cheesecake as you run past course assistants.

Nonetheless, this isn't for everyone, as coordination of simultaneous cheesecake consumption and marathon running can frankly be tricky. It gets tougher when combined with the need, continuously, to solve simultaneously for m and n using:

$$\Delta\,\text{months} = [J_{11}]\,m + [J_{12}]\,n \tag{4.3}$$

$$\Delta\,\text{pounds} = [J_{21}]\,m + [J_{22}]\,n$$

where "**Δ months**" is the difference in months between your desired longevity and the readout on your wrist, and where "**Δ pounds**" is the difference between your desired weight and current weight as registered with your instrumented running shoes. The elements $\mathbf{J_{ij}}$ correspond with the Jacobian elements of Eq. 4.1, e.g. $\mathbf{J_{21}}$= **-5 pounds/1 marathon**.

Finally, you need a control law. This is a rule to guide your running effort in accordance with the current number that solving Eqs. 4.3 gives you for m, and your eating effort in accordance with the same equations' requirements for n. Often simple proportional control laws are used: Running effort = $\mathbf{K_m m}$; Eating effort = $\mathbf{K_n n}$, where $\mathbf{K_m}$ and $\mathbf{K_n}$ are positive, constant "gains" that are judiciously chosen. "What exactly," you may be saying "do you mean by 'effort'?"

Herein lies the beauty of servoing. Provided your definition of "effort" includes stopping running when $\mathbf{K_m m}$ goes to zero and stopping eating when $\mathbf{K_n n}$ goes to zero, the use of feedback produces effective and easily stabilized performance for a range of definitions of actuator input effort. These may entail applying pedometers to measure your speed, thereby defining "running effort" directly in terms of how fast you run; or you might apply some other definition. The same goes for visual servoing. Academic papers have demonstrated using impressive notation that desirable closed-loop performance characteristics can be had if the control-law output for visual servoing is commanded joint-rotation velocity or some more primitive motor "urge", such as voltage to the actuators. This is especially true if integral and derivative terms are added to the control law; see Chapter 2. With differential equations involved in the action, such as with the joint servomechanisms described in Chapter 2, issues of closed loop stability and coupling between the action of the joint level control and the Jacobian based visual feedback may require some analysis and gain adjustment. (Considerations of such coupling and its effects upon performance, stability and overshoot mostly disappear with the half step, start and stop method suggested for our pan/tilt control.) But for the heavier robot arm visual servoing generally requires continuous motion. Adding somewhat to this burden of closed loop stability is the issue of finite time delay for processing images. Such delay is also instability-inducing and requires analysis.

You are nevertheless optimistic that continuous cheesecake eating and marathon running, taken with similarly judicious control laws, will allow you to reach your simultaneous goals of longevity and weight gain. So you press ahead.

Until the sky falls in. There she is: Dr. Anne in a "perp walk". It seems that she has been selling guaranteed lifetime incomes based upon fraudulent assurances of longevity. And she turns out not even to be a real doctor. (Who can you trust anymore?) Oh yes, there's a minor charge of selling bogus diagnostic kits and equipment.

So there you have it: a brilliant idea lost for lack of good feedback sampling. And the same holds true for visual servoing. Just as we have no instruments to measure longevity, so too the ability to sense the remaining distance to closure - real time in an image - of two joining parts (excepting a few possible instances such as needle threading) is prohibitively difficult to engineer - even though, clearly, people do it all the time.

In some ways it's like the human ability to recognize individuals who they've seen in the past from recent pictures of their faces, despite wide variations in the acquired images' incident lighting, or the individual's position relative to the camera, or his or her facial expression. The ability just can't be reproduced with computer-vision software. We wish it could.

Interestingly, both these unmatchable visual abilities have important survival value and may be exceptional vis-à-vis other categories of human image understanding. Evidently (*Economist*, October 2004) the ability to recognize individual human faces is advanced well beyond the ability to discern, say, different automobile models. This difference is due presumably to the value to survival of distinguishing readily among individuals of one's own species. Likewise, some evolutionists believe that humans' rapid brain development directly corresponded to, and developed synergistically with, the ascendancy of the highly advantageous crafting and use of hand tools - with the attendant need to discern visually the ongoing proximity relationships between manipulated objects.

As running and eating produce myriad effects on all kinds of bodily outputs and hence sensors that monitor them, so too "closure", the approach of one object surface toward another, results in complex shadows and interplay of light reflecting between the proximate bodies and into cameras. Extracting the needed information from images in order to feed back the "error" in real time is fraught with problems, too many problems to for real-world use.

Our spot of laser light does not carry with it this problem because closure onto its target does not affect the spot image. There is no concomitant encroachment into the image of a physical body. And image differencing can be applied for fast and robust image analysis because the key cue is the structured point of light itself - at least provided the surface on which the spot lands has a reasonably large radius of curvature compared to the size of the spot. Hence, with robustness of convergence to the aforementioned large disparities possible between the Jacobian one actually applies and the real-world joint-movement-to-camera-space-response ratios, it is ideal for this key element of point-and-click human supervisory control.

Behavior based robotics

Humans must once have survived for long periods of time in the wilderness. This seems certainly to have required on our part awareness, skill, robustness, and the underrated "common sense". But all sorts of animals, many of which appear to have only, of these four seemingly necessary attributes, "robustness" going for them, also succeed at the same game. It's hard to know just what goes on, for example, in the lima-bean-sized brain of an alligator, but judging by its behavior - it will snap at and try to devour anything close-by that moves - the imperative of getting food does not place much demand on it in terms of what we might call object recognition. Yet these creatures have survived in their present form millions of years in difficult settings; many have even been known to get by in New York City sewers: lots of success; little cognition; and no doubt fully autonomous. Less overpowering species have similarly thrived, often, such as with bees and ants, using behavior that is cooperative though certainly not in a reasoned or premeditated way. Again, judiciously programmed responses to sensory input - like the imprinting of ducks to follow the first animal they see after being hatched.

Could it be that image understanding, or other forms of gaining a qualitative/quantitative read on the immediate environment, is overrated as a prerequisite for successful robot behavior? In fact, some of the most useful and/or amazing machines that have neared or in some cases easily crossed the threshold of Generally Accepted as Robots (GAaR) work in this way: One or more actuators respond to one or more sensors in a programmed-in, perfunctory way. There is no analysis or image recognition or calculation, really; at most there is just a decision tree: if sensor three registers a reading above six while sensors two and four remain below nine apply four more volts to actuator four. That sort of thing. Determined reductionists might argue that any algorithm-based system, no matter how much it depends upon apriori science and mathematics and modeling for sensory processing, is essentially no different from this basic actuator response to sensory input. That indeed what humans do is just an elaborate form of the same thing. (We may be able to dismiss such criticism as a deterministic response to this page.) Nevertheless, advocates of behavior-based robotics look more for pragmatic sensor-input-to-actuator-output relationships that happen to produce a desired overall behavior in domains that typify task settings of the required application. This would contrast with objective high-level system "understanding" of the objects with which the robot must interact.

Consider swimming-pool-cleaning machines. These have actually been around for decades. They autonomously clean the sides of swimming pools, moving along their walls much as a snail might move along the side of an aquarium. The rule they follow is simple: "If the edge or end of the wall is encountered change direction." With time the odds increasingly favor that the device will have covered every square inch of the pool at least once.

A very successful use of this approach are pet dogs and other moving mechanisms that respond to various sensory stimuli in ways that evoke instinctive human sympathy. These show the ability of machines to connect with humans on an emotional level. The sequential responses of the mechanisms also show the ability to approach complex interaction with (in this case human) environmental entities that are made more elaborate, subtle and efficacious with complexity of the programmed-in, tweaked-(evolved)-over-time, logic.

Another interesting possibility for inclusion as an example of behavior-based robotics this is Honda's famous *Asimo* robot. This balancing machine can shift weight from leg to leg, moving forward as it does; and it has other anthropomorphic moves such as dancing. It might be thought that such feats are heavy on Newtonian dynamics, with lots of numerical integration of nonlinear equations of motion to determine needed actuation at any moment; and indeed we presume the system's responses are consistent with classical mechanics. But developers took a more empirical, trial-and-error approach to the set of rules that in fact governs actuators' responses to sensor input. *All with an eye to bringing about acceptable overall behavior.*

The most salient example of behavior-based robotics to many is the *Roomba* vacuum cleaner. This small, flying-saucer-shaped, wheeled device moves completely autonomously in typical rooms within a home. Evolved over time, recent examples detect and steer into dirt where it is found. Other programmed-in rules include responses to move away from walls and floor discontinuities such as downward staircases. Comprehensiveness of floor cleaning is, as with the swimming-pool cleaner, left to the laws of chance; but the device has won devotees worldwide for its effectiveness, convenience and novelty.

With examples such as these it doesn't seem such a stretch to think that behavior-based robotics might comparatively soon produce all kinds of domestic and other mechanical helpers. Elusive problems of image analysis disappear. Existence proofs from nature of a broad range of effective behaviors abound. And really the line of inquiry has only just begun. Optimism, in fact, for this broad track of thinking made behavior-based robotics probably the main robotics research theme of the late nineties.

In 2003 the authors attended a supercomputer symposium in Phoenix, Arizona. There we showed the automatically guided wheelchair system discussed in Chapter 10. The autonomous chair executed a repertoire of six or seven trajectories - previously "taught" paths that connected junctures of interest, specific poses in the plane of the floor, within a very constrained set of walls. A disabled rider unable to steer his power chair directly, could realize a high degree of autonomy within even a tightly constrained home by simply choosing destination sequences. Visitors were invited to do so in Phoenix, selecting destinations from a menu using a chin switch.

A physicist who had read about behavior-based robotics stopped by to chat. When the discussion quickly turned to behavior-based robotics I expressed doubt that there would turn out to be a large number of useful purposes that could be served with systems so designed. "Well what would you want a system to do?" he asked. "How about transport a disabled wheelchair rider from a current location in his home to one that he specifies?" I said. The physicist thought. Only the most persuaded Procrustean could ask a rider who commanded transport, say, to the bathroom, to await a random-walk delivery! Neither of us was sure how this task could be addressed with the behavior-based approach.

And that may be the point. The size of the universe of possibilities is very difficult to judge. The imagination simply fails. Natural systems may not, even if we knew how they actually worked, have counterparts that one would pay to automate. And it isn't clear how much additional competency could be added practically to early manifestations of behavior-based robotics in existence today. Then there is the issue of reliability, and reliability assessment.

In contrast to this, we believe that there is a reasonably clear path ahead from the technologies given in this book to a range of plausible uses. Laying this out is the purpose of the rest of the volume.

CHAPTER 5

THE PHENOMENAL ABILITY OF HUMANS AND OTHER ORGANISMS AND IMAGE FORMATION

The "existence proof" that natural sight can be the basis for useful control of natural and engineered mechanisms, despite the vagaries of electromagnetic energy reflected off a surface and registered on a plane, makes us (so far way too) optimistic for robotics.

The natural ability to recognize the objects in an image seems to become more remarkable the more effort is placed on attempts to do it artificially. Take for example a digital picture of a cellphone and a pair of eyeglasses resting on an otherwise empty tray - the example of Chapter 1. Besides the actual surface geometry of the objects themselves, a wide assortment of lighting and reflection characteristics factor into the "grayscale data" that comprise the image (think for example of the range of gray tones in a black and white photo.) For a very wide range of these variables most people would recognize the objects instantly. Perhaps most stunning about this ability is its qualification as a very reliable solution to a chailenging instance of the "inverse problem," a category of problem associated with everything from crime solving to paleontology to cosmology to nondestructive evaluation of internal flaws in an aircraft.

In a former (technical) life, late in the 1970s, the writer worked in the nuclear-power industry. Part of the job involved demonstrating the fracture integrity of reactor vessels - those very large high-pressure metal boilers which contain nuclear reactions at their core. The fear was failure - a catastrophic rupture due to the mix of high pressure with hidden, internal flaws in the vessel and a weakening of the metal due to a constant neutron bombardment from the core reaction.

A science had been developed to test the metal for the presence of internal flaws. Emitters of high-frequency sound were placed at one or more known locations on the (cooled) reactor vessel surface. Detectors of the response at one or more suface points were placed at other, known surfacc locations. The science of the acoustics of sound-wave travel through a known metallic medium resulted in computer models that could predict how any given internal flaw or crack would affect the measured responses. The predictions were accurate and exact. If geometric characteristics of an elliptical void, say one inch by three inches in size, buried at a known location and orientation inside the wide girth of reactor vessel wall, were input into the computer program, the program would forecast exactly what effect this flaw would have on the acoustical readings from a kind of microphone placed on the surface, in response to emitted sounds. Amazing stuff.

It's tempting to suppose that such a capability would solve the problem of demonstrating safety. Just measure the sound output, go to the computer model, and there you would have it; you'd be able to "see", or at least infer, all internal flaw geometries. There is, however, a subtlety, and it has to do with this being another example of the "inverse problem".

While the "forward problem" (given a flaw, determine the ultrasound-sensor response) had been solved, our interest was a bit different: Given the ultrasound response what are the actual internal geometries of the flaws and voids that influenced that response? This latter question could actually be impossible to answer, even if the computer program's solution to the forward problem were perfect. It seems that for any given input-output pairing, the computer could tell you (if there was time to run enough trial cases) more than one, more than ten, more than a thousand internal flaws or cracks - or combinations of internal flaws and cracks - each of which would result in exactly the same ultrasound samples, given exactly the same controlled sound emissions.

What to do? In order to make use then of the experimental information, it became necessary to impose constraints on the interpretation of the data, in some way to limit the possibilities of just how any causal flaws might configure themselves inside the massive vessel wall. Fortunately, the regulatory powers were less concerned that this interprepretation was "right" - characterizing geometrically the actual internal imperfections - than that it was "conservative". The actual question therefore became answerable: "what is the most hazardous configuration of flaws that would give rise to the particular data acquired?" If this most hazardous interpretation was compatible with safe operation then all was well, and the vessel could continue to operate.

If it looks like a duck ...

In the light-based-imaging game, the "forward problem" is likewise substantially solved. Three-dimensional animated movies are an example. They begin with a data base of a virtual three-dimensional "reality", say the furnishings in a child's bedroom. Added to this computer information is the location, intensity, wavelengths and other information pertaining to the virtual lighting that is to illuminate this scene. A virtual camera location and lens characteristics are added, as are color and other reflective characteristics of the surfaces, and the program can create an amazingly lifelike two-dimensional scene. "Move" the virtual camera a bit and the same room is seen from a different perspective.

There are some useful analogies between the ultrasound computer program discussed above and the software used to create two-dimensional pictures from a database of the three-dimensional world. Consider the virtual lighting of the *Pixar* movie. This is a specific, presumed source of energy that interacts with the "scene". The computer program calculates how this interaction, given the database of the geometric and reflective surfaces in the bedroom, will cause light to appear in the virtual camera. This can be taken to be analogous to the waves of sound incident on the reactor vessel wall. The program used for nondestructive evaluation of internal flaws calculates how this energy, propagated through a solid, elastic medium, will interact with the mathematically modeled internal flaws and reactor-vessel materials and external geometry. The virtual camera and lens of the Pixar movie would be like the ultrasonic receiver or microphone of the nondestructive evaluation. The remarkable thing is that the analogy extends to this: If it could be run enough times, the same "forward-problem" computer program that created the movie scene could find many more configurations of surfaces (i.e. completely different "stuff" that might be inside the child's room) which combined even with the same virtual lighting and camera positions would produce exactly the same images. (One of these possibilities would be if the room contained a flat, very exact, painting of the intended furnishings.) Add to this the fact that movie viewers - unlike nuclear engineers with their ultrasound problem – have no idea what incident light or camera-sensor characteristics went into the scene creation, and it seems quite remarkable that the audience's interpretation of the objects in the scene is essentially correct. *Pixar* does the forward problem, using kajillions of computational, arithmetic operations and lots of lead time, but the audience does the much harder inverse problem practically instantly! There's no "most conservative interpretation" that this audience has to rely on, as with the nuclear-reactor demonstration of safety; we, the audience, simply "get it" – every time, or close to it – and instantly. And we can do the same with real photographic images of real scenes, or with direct visual access to the actual world.

That "seeing is believing" is a tribute to just what good guessing machines we humans are.

How do we manage such accurate, instantaneous solution to a difficult inverse problem? In some sense, templates or constraints on possible interpretation are needed. Such templates must be in us somewhere; how else could we "see" Dudley Dooright while gazing up at a cloud formation? Or the Virgin Mary on a piece of cheese toast. But the underlying templates and interconnected constraints are subtle. They seem to improve with experience. And they cannot currently be approached in effect with artificial image analysis.

As a pragmatic matter, therefore, the application of the human ability to direct mechanical dexterity is a short-term must for most of the objects of manipulation that lie outside of highly constrained situations such as factories. Ongoing "human in the loop" control, however, is not necessary, we believe. This is fortunate because ongoing human control of mechanisms is very imperfect and/or very slow. Between the time of the writing of the introductory chapter, for example, and the present writing, NASA made the decision that human in the loop control of Canada's Special Purpose Dexterous Manipulator, called "Dextre", was not likely to be reliable enough to save the Hubble telescope.

A thousand points of light

Yet the human ability to select surface points within a presented image is potentially much better than our ongoing ability to provide real-time guidance and control. Moreover, as discussed in Chapter 8, a vast range of real-world task objectives can be defined mathematically in terms of the desired geometrical movement of a three-dimensional end effector with respect to surface points so selected. If counterparts to image-plane locations of these points are conveyed using sometimes one, sometimes three, better still thousands of points of laser light, controlled autonomously on-site, then the robust method of Camera-Space Manipulation can be used to achieve a level of precision and ongoing steadiness of the "pure motion" of the manipulator that will unleash a vast array of possibilities for exploiting the human-mechanism partnership.

To see how this works, a few words are in order about the way in which light actually creates images in a charged couple device camera. Consider light incident on a flat surface, say a wall. Suppose (for this immediate discussion) that the light is uniformly intense across some region of the wall. "Power" of the reflected light off a given small region of the wall, for example a square millimeter, could be measured in Watts per steradian, where a steradian is unit of "solid angle" (think for example of the cone-shaped portion of a funnel).

Depending upon the surface characteristics of the wall, this power will vary. If for example the wall is black, the reflected-light power is nearly zero. Additionally, the power can vary depending upon the orientation of a small solid angle of reflection with respect both to the wall and the source of illumination. If the surface is shiny, then the power is much greater per steradian in directions where angle of reflection is close to angle of incidence. If the surface is "flat" (in the sense we use when speaking of a flat paint, for example), or "Lambertian," then the power is nearly the same in every direction with respect to the wall orientation and the incident source's direction.

Whatever the reflective characteristics of the wall, it is the business of a lens acquiring an image of the wall to gather all of the light within the number of steradians encompassed by its lens opening and concentrate this energy onto a single small region on the Charged Couple Device. If it succeeds at doing this we say that the image is in focus. The reason that light energy from a finite number of steradians must be accumulated has to do with the need for finite rather than infinitessimal power - as naturally required to "register" with the sensing charged-couple device. In fact that need for a finite amount of energy necessitates not just gathering of light from a finite-sized piece of solid-angle steradians but also accumulating light energy over a finite period of time (power being energy per unit time). Think of a double integral in calculus.

This ability to gather light reflected in a range of directions – i.e. over finite steradians – off the same small, light-reflecting region of wall is thanks to lenses, and their ability to take advantage of the characteristic of light to change direction at junctures of transparent-medium-density change, such as when the light crosses the threshold from air to glass.

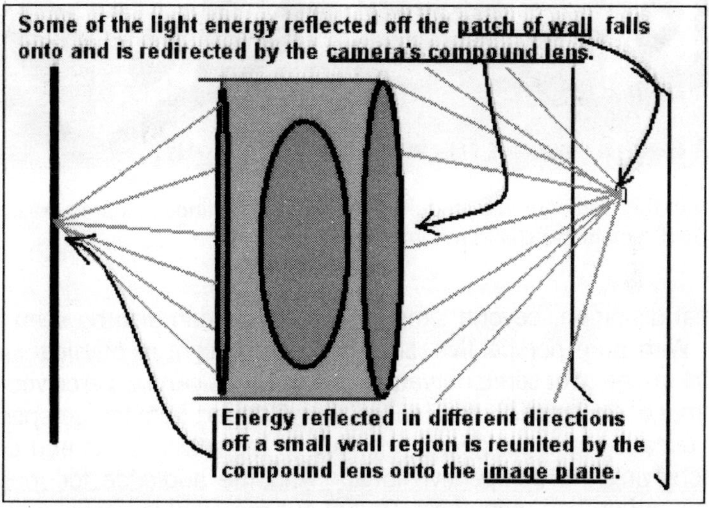

Figure 5-1. The job of the compound lens, designed using yet another "forward problem" computer program, is to focus a significant number of steradians of the light reflected off the small patch of wall onto a very small region of the image plane.

Just as computer programs exist to determine how light will reflect off surfaces, so too computer programs exist to determine the effect associated with refraction. Such programs are used to design compound lenses that accomplish the focusing of light as per *Figure 5-1*. Each design that results in adequate focus and light accumulation, however, will distribute the light from a given juncture on the physical surface into the image plane in a slightly different way. The resulting distortion can be obvious, as for example with the use of a very wide angle lens. In the two pictures of a block of *Figure 5-2* the block appears to be the same size in both images. Both of the two lenses focus the same light reflected off the same object onto the same image plane, but the result is apparently different. In the image on the left, the picture is taken from some distance away using a long-focal-axis lens. As a consequence, the image has characteristics of an orthographic mapping, that is a straight or flat projection of points in three-dimensional physical space onto a plane. The image on the right, on the other hand, exhibits more of the perspective effect: lines that are parallel in physical space would intersect if their image in camera space were extended.

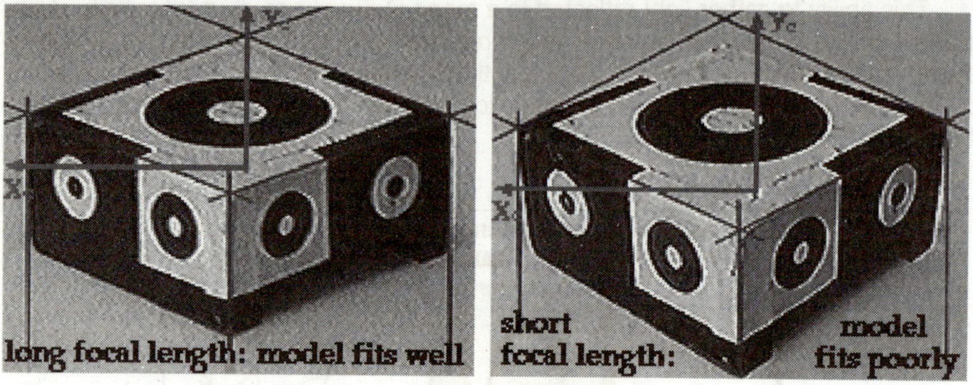

Figure 5-2. With the "orthographic model" of lens mapping, lines that are parallel in real, physical space would remain parallel in the image.

Additional distortion, beyond simple perspective, can also be seen in the rightmost figure. With pure perspective, lines that are straight in physical space remain straight in the image. Yet some curvature of edge lines can be perceived in the rightmost block image. In practice, real lenses always depart from the perspective or "pinhole" ideal because lens designers' priorities iie more with focus and depth of field than with achieving the perspective ideal. And the audience for most images is human beings rather than machines. Part of our remarkable image-analysis ability is that we humans are not fooled by even fairly high lens distortion. We still interpret the image information correctly. We may not generally notice distortion in fact unless it is pointed out.

With computer vision we must be conscious of these effects. Fortunately, a few things work in our favor. One is that, even for wide-angle lenses where there may be plenty of overall distortion, the mapping of objects from a small region of physical space is orthographic. For instance, referring to the two images of *Figure 5-2,* in addition to the left image's characteristic of parallel lines remaining parallel, the circular shapes of the physically round cues map into ellipses in that image. That is an orthographic effect. Moreover, the ratio of light to dark to light as one sweeps through the center of that image is the same as with the original surface (see Appendix C). This too is a property of an orthographic mapping. If we examine the more distorted right image, the larger cue's appearance departs significantly from this orthgraphic ideal. The smaller cues, however, of the same image nearly conform to the ideal. That allows us to apply various geometric invariants of an orthographic mapping to identification of the cues, as discussed in Appendix C, without the need to worry about global departures of the actual lens from this ideal - provided the cues occupy a small region of the overall image. Additionally, provided the spots of light from a laser pointer are small with respect to the overall image and provided the radius of curvature of the physical surface upon which the spots fall is large compared to the spot's radius, another orthographic advantage holds: The spot's center will be found in all cameras with visual access to the spot in such a way as to represent the same location or juncture on that physical, three-dimensional surface.

Just as confinement to a relatively small region of physical space results in a better representation of the camera/lens mapping of that region by an orthographic model, so too an even lesser confinement results in better conformity with the perspective or pinhole ideal. These "asymptotic limits", as discussed at length in Chapter 7, are important for the implementation of camera-space manipulation.

CHAPTER 6

CAMERA SPACE MANIPULATION: THINGS COMPUTERS AND MACHINES DO WELL; THE SIMULTANEOUS USE OF MULTIPLE POINTS OF VIEW

We humans successfully apply our dexterity to a very wide range of useful ends. But with the stubborn limitation seen in Chapter 3 regarding the applicability and usefulness to date of artificial mechanical dexterity it becomes difficult to imagine the broad range of application that effective control of such mechanisms could bring about. Subjecting this dexterity, however, to the power of computation using the technology herein will, we think, open new prospect that may dwarf in effect the productivity and quality enhancements that other applications of computers have brought.

In the introductory chapter it was noted that while humans and robots both may possess mechanical dexterity, the experience to date is that usefully versatile control of this feature overwhelmingly favors the humans. Yet, strictly from the point of view of their mechanical attributes, putting aside the wide present-day gulf that separates the two systems' real-world compatibility with visual guidance, there are some clear advantages of the machines: Even today they can be built lighter, stronger, faster, and steadier than any human. The angular positioning (see Chapter 2) of each joint can be resolved and controlled to within thousandths of a degree. They don't tire, and diminution of performance tends to be gradual and monotonic over time. What is the culprit in our inability to exploit such capability? Control. We simply aren't good at controlling the robotic arms' rotational servomechanisms in such a way as to bring about real-world-useful, real-world-robust, real-world-precise action of the manipulated tool or body as needed to accomplish the range of ends for which robots would otherwise be well suited. It isn't that we can't control each degree of freedom's motor to track a computer-specified angular trajectory; that part is exceedingly accurate. The problem is getting the computer to know what sequence of rotation, orchestrated among perhaps six robot joints, will correspond with moving the manipulated tool or other body relative to the workpiece to achieve the manipulation goal.

Consider the advancement beyond human dexterity that came about from the use of what amounts to a two degree of freedom robot: the x-y pen plotter, as indicated in *Figure 6-1*. Such plotters were widely used before the days of laser printers, and are still used today for making certain types of plots and drawings. The mechanism basically moves a pen tip across a page in accordance with an x-y sequence of points. The performance of such a system when driven by a computer is awesome. No human could accurately draw a cubic spline for instance, mathematically specified, onto a blank page. Yet the pen-plot robot does this and many other precise representations with a degree of perfection limited only by its own calibration. Can we extend this magnificent versatility and precision to other robots operating in the *three*-dimensional world? A six-degree-of-freedom robot has all the versatility of movement needed to transfer via the plotter's pen a computer-generated shape onto, say, the surface of the ostrich egg illustrated in *Figure 6-2*. The pen's pressure must be firm enough to leave a mark but not so firm as to damage egg or pen tip. If the pen orientation is

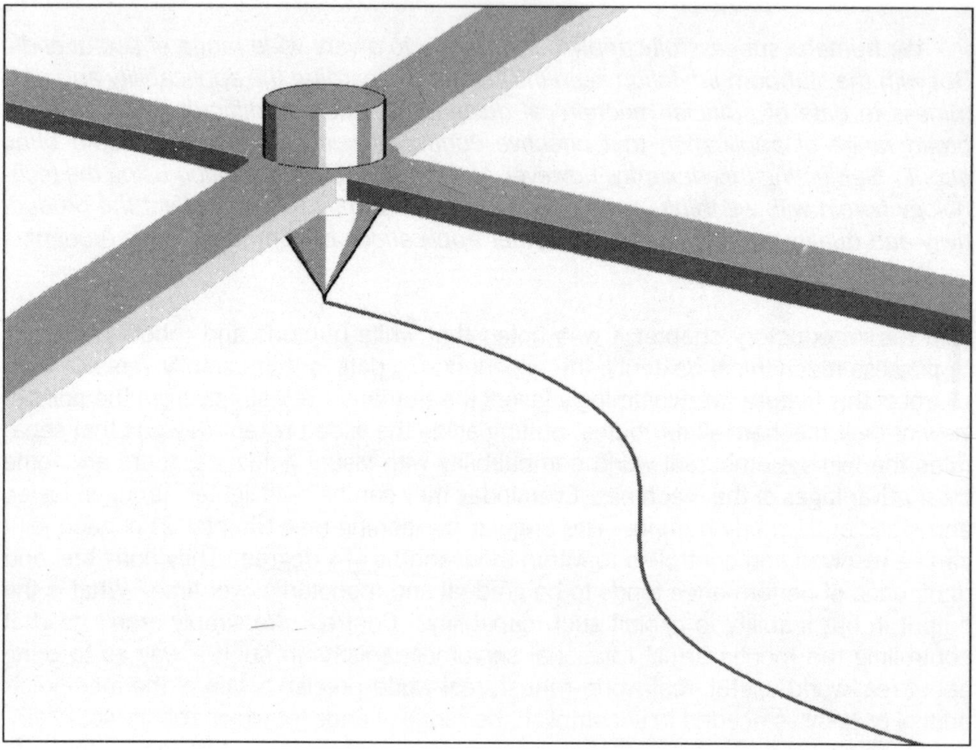

Figure 6-1. A pen plotter is like a two-degree-of-freedom, calibrated robot. The simplest shapes such as a circle of prescribed radius could not be created by a skilled human using hand-eye coordination alone, without tools like a compass. The computer, however, allows for the specification of shapes and forms well beyond those for which hand-drawing tools even exist. And provided it sustains good calibration the ***x-y*** plotter can manifest these shapes.

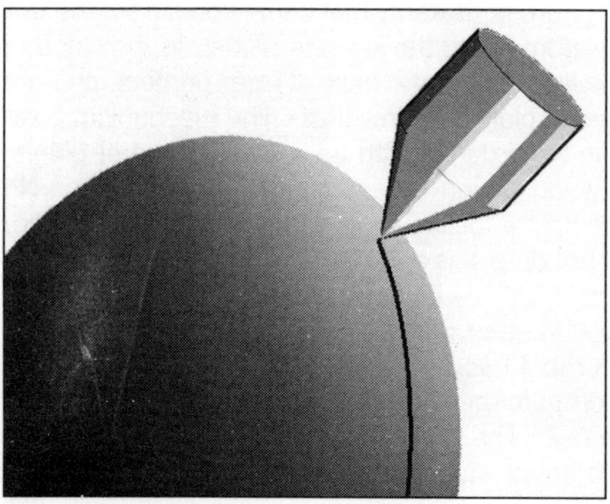

Figure 6-2. The ability of camera-space manipulation to bring about dexterous robot control in three dimensions relative to real, arbitrary surfaces is illustrated with the drawing of a computer-generated shape onto an ostrich egg.

reasonably perpendicular to the egg's surface at any point of contact (a condition we will ensure later) this force restriction might translate, with the actual compliance or "springiness" of egg, pen tip and mechanism, to a requirement of tip-to-egg-surface positioning-control precision within 0.25 millimeter. With even robot manufacturers' claims of their robotic mechanisms' global accuracy ten or more times worse than that, and with all of the vagaries of real optics and camera electronics, how is such precision to be ensured?

For illustration, suppose that the starting egg-surface point is selected by a remote user, over the internet, with point and click. Presented with a high-resolution image from arbitrarily placed "Camera 1", as indicated in *Figure 6-3*, the human supervisor employs a graphical user interface (e.g. with a mouse) to choose this juncture. Then an autonomous pan/tilt unit directs a narrow-beam laser pointer using the robust means of convergence detailed in Chapter 4 and *Appendix D*. Automatic pan/tilt motion proceeds until the laser spot falls on the physical juncture corresponding with the *Figure 6-3* selection from the image representation used by the remote supervisor.

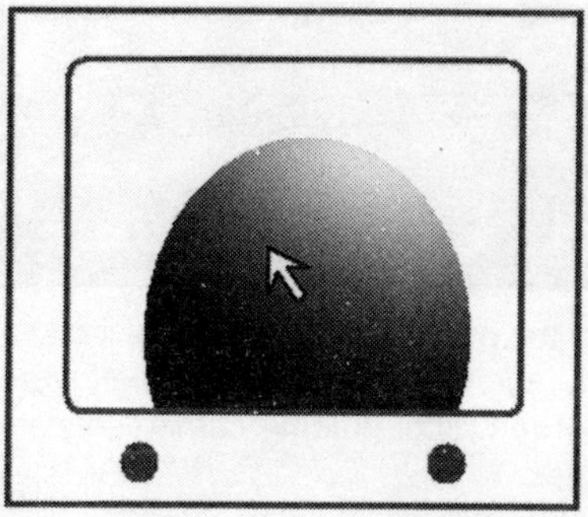

Figure 6-3. Presented with an image of the workpiece, one that is acquired from Camera 1, a remote supervisor selects the surface point on which the diagram is to begin, using computer-mouse point and click.

Figure 6-4 shows in its upper-left corner the terminal state of the laser pointer. The center of spot illumination on the egg can be made arbitrarily close to the user-selected point of *Figure 6-3*. Once this terminus is reached camera acquisition of the spot can be synchronized autonomously with laser-pointer on/off control in such a way as to facilitate the acquisition in all participating image planes or camera spaces of a "differenced image". A differenced image is actually a subtraction of one image's matrix of grayscale values, acquired with the laser pointer turned off, from a second matrix, acquired with the laser turned on. Think of the lightness of each picture element or "pixel" as being proportional to a number in each picture's grayscale matrix. The net result – the differenced image – is a grayscale matrix that leaves mostly the effect of the laser spot alone. Appendix D gives one of several possible highly reliable and robust means by which a computer can then locate the spot center in each participant camera: Provided the spot's radius is small compared to the surface point's

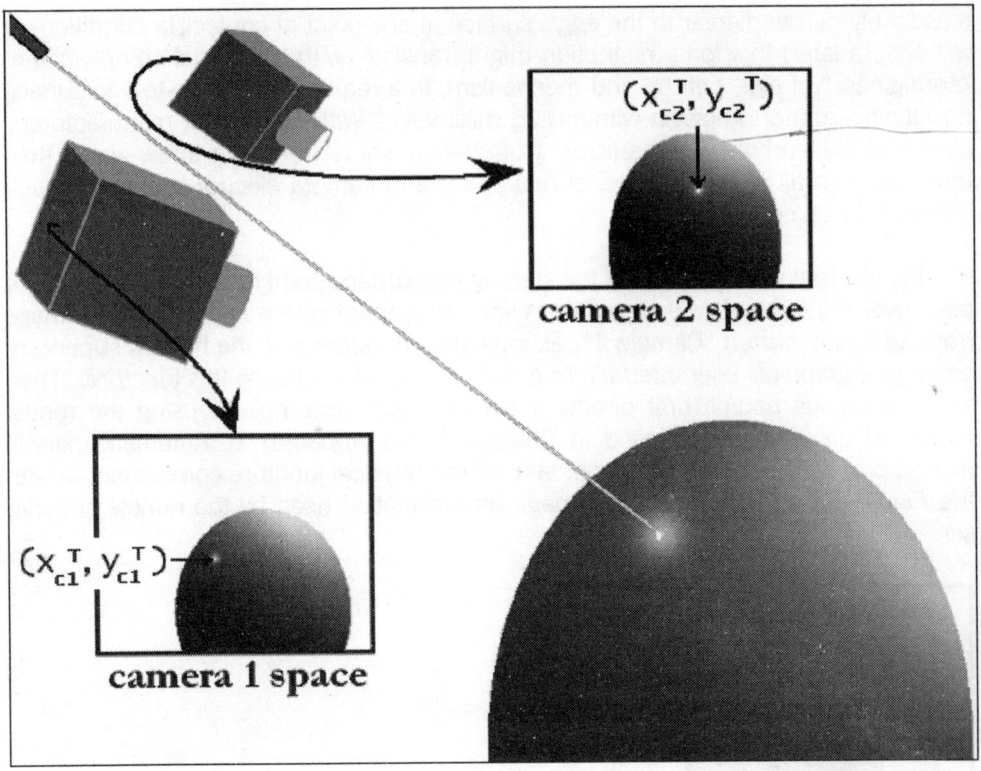

Figure 6-4. The laser pointer (upper left) comes to rest after the pan/tilt unit that directs it causes the spot center to fall directly on the user-selected juncture in Camera-1 space. See Figure 6-3. Image differencing is then used to ensure that both of the robot-guiding cameras locate the same target surface juncture in their 2D reference frames.

smallest radius of surface curvature, a simple "mask" applied to each camera's differenced image assures that the supervisor-selected physical juncture is found in each camera's two-dimensional image plane or camera space. We thus identify in Camera 1 a target camera-space position (x_{c1}^t, y_{c1}^t), and in Camera-2 a target camera-space position (x_{c2}^t, y_{c2}^t), as indicated in *Figure 6-4.* Note that this procedure could be extended to any number of additional participant cameras with a good view of the surface. From a surface-point selection using the single Camera 1, any number of additional observing cameras can use the image-differencing/spot-center-finding approach to locate within their own camera spaces the user-selected point. And any of these cameras could be designated the "Camera 1" – what we call herein the "selection camera" – whose image is used to select the target point. Somewhat counterintuitively, selection of this point in the reference frame of the two-dimensional computer monitor is sufficient to specify a *three*-dimensional maneuver objective. The reason for this is that the physical egg surface itself provides the "constraint" that makes selection from the two-dimensional Camera-1 space correspond with a unique point in the three-dimensional world.

Suppose now that the robot guides the pen's approach to the selected surface juncture with an orientation that is nearly perpendicular to the surface at the targeted point. Suppose further that the procedure outlined below succeeds in identifying, and servomechanisms of Chapter 2 succeed in driving, the robot's internal joint rotations such that the white X of *Figure 6-5* would be precisely placed, absent resistance from the shell itself, at the user-selected juncture on the egg's surface. Certainly, the portion of the tip in front of *Figure 6-5* white X can't occupy the same physical space as the egg shell, so something will have had to give. But with the small, half-mm interval selected for defining the white X's location with respect to the tip's very end, the magnitude of this "give" is small; the 1/2 mm was actually picked precisely with this yielding of robot and egg surface in mind. The compliance of shell, robot and pen tip, working together in resisting intrusion of the tip, is fairly predictable. In other words, assuming the shell is adequately supported and will not displace or shift when a small force is applied, the approximate force of resistance that should result from achieving the robot's internal configuration as planned absent such resistance is maybe two ounces. The needed relationship between overall yielding or "give" and force of contact is typically knowable with pretty good accuracy – if only the needed precision of relative positioning can be achieved. The half mm of total "give" required by this system if servomechanism command of the robot's joints were such that the white X would, absent impediment, be located in space as discussed above would result in just about the ideal force of contact for leaving a good mark. Even if we miss this robot joint pose to the extent of the white X's unresisted position being off in any direction by up to 0.25 mm relative to the undeformed shell's surface point, the pressure will still be within the pen's "sweet spot" or range of tip force needed to deliver a good mark.

Figure 6-5. If it could be seen, the juncture on the pen tip represented by the white X would move about in camera space as a function of the robot's current joint angles $\theta_1, \theta_2 \ldots \theta_6$. Suppose that the functional form of this relationship were precisely known.

Computation of the internal robot angles, θ_1, θ_2... θ_6, needed to deliver the selected pen-tip juncture to the shell-surface location of the laser spot occurs as follows.

Using an approach described in detail in the next chapter, the controlling computer has available key algebraic functions: One of these is the position of *Figure 6-5's* juncture X on the pen tip, *in the two-dimensional reference frame of Camera 1*, as a function of the robot's joint rotations. Call the Camera-1 x-component of this function $f_{1x}(\theta_1, \theta_2... \theta_6)$ and the Camera-1 y-component $f_{1y}(\theta_1, \theta_2... \theta_6)$. With these two functions known, one could specify any combination of six joint angles and know where in a Camera-1 image, if it could be seen, the white X of *Figure 6-5* would be located. Called herein the "camera-space kinematics" for the robot in Camera 1, these functions have a counterpart in mainstream robotics: A robot's "forward kinematics" could be used to describe the coordinates of a selected point on its end member, even a hidden point such as our white X, as a function of the internal angles of the robot. Being a physical-space description, the forward kinematics model would entail three components, rather than our camera space's two components, of position. These three components are usually given with respect to a physical reference frame that is stationary relative to the robot's base.

Two important questions: **1.** How could camera-space kinematic functions f_{1x}, f_{1y} be knowable with a sufficiently high level of precision for our 0.25 mm of maximum error when the best robots' forward kinematic relationships are not globally accurate to within even a centimeter? **2.** If the camera-space kinematics *were* known with adequate precision, how would they help us calculate the joint rotations θ_1, θ_2 ... θ_6 to command for the robot's six positional servomechanisms in order to place the tip as required relative to the egg surface?

Let's consider this second question first. Begin by assuming that, in addition to knowledge of Camera 1's $f_{1x}(\theta_1, \theta_2... \theta_6)$ and $f_{1y}(\theta_1, \theta_2... \theta_6)$, we know similar functions $f_{2x}(\theta_1, \theta_2... \theta_6)$ and $f_{2y}(\theta_1, \theta_2... \theta_6)$ for Camera 2 of *Figure 6-4.*

Let's hold fixed for the moment the three wrist angles θ_4, θ_5, θ_6 of our robot of *Figure 2-2.* Note that three-dimensional *position* of our white X, and *orientation* of our tool holder, will both generally depend on *all* six angles. For now, however, we'll settle for only approximate pen perpendicularity upon "arrival", and we take this to be safely assumed with judicious selection of fixed $\theta_4=\theta_{40}$, $\theta_5=\theta_{50}$, $\theta_6=\theta_{60}$. Later in the present chapter, full-pose calculation as required to achieve complete camera-determined position/orientation with full coupling in all six angles θ_1, θ_2... θ_6 is discussed.

That leaves θ_1, θ_2, θ_3 to be calculated – and this part must be highly precise – such that positioning of the white X occurs at the center of the converged laser spot on the egg surface i.e. the target point of interest, to within a quarter of a millimeter. Recall we are assuming precisely known camera-space kinematic relationships $x_{c1}= f_{1x}(\theta_1,\theta_2,\theta_3,\theta_{4o},\theta_{5o},\theta_{6o})$, $y_{c1}= f_{1y}(\theta_1,\theta_2,\theta_3,\theta_{4o},\theta_{5o},\theta_{6o})$, $x_{c2}= f_{2x}(\theta_1,\theta_2,\theta_3,\theta_{4o},\theta_{5o},\theta_{6o})$, $y_{c2}= f_{2y}(\theta_1, \theta_2, \theta_3, \theta_{4o}, \theta_{5o}, \theta_{6o})$, where (x_{c1}, y_{c1}) are the coordinates of the white X in Camera 1, absent resistance of contact – if our pen-tip point X could actually be seen; and (x_{c2}, y_{c2}) are the coordinates of the white X in Camera 2. Referring to *Figure 6-4*, the laser pointer has been applied already such that camera-space targets (x_{c1}^t, y_{c1}^t) and (x_{c2}^t, y_{c2}^t) have been identified precisely and consistently with the human supervisor's selection.

Sufficient camera-space conditions for placement success

By and large it is safe to assume that light-frequency differences don't affect lens mapping. In other words green light and red light, say, reflected off the same surface point, show up in virtually the same location in our black-and-white, in-focus image. Also, as mentioned, we assume the egg is well supported, and does not slip with small force of contact. A sufficient condition for achieving the maneuver goal, then, will be simply to control the robot such that $(x_{c1}, y_{c1})=(x_{c1}^t, y_{c1}^t)$ and $(x_{c2}, y_{c2})=(x_{c2}^t, y_{c2}^t)$. In other words placing the white X in the right place in each of our two camera spaces will correspond with placing it as desired in physical space. A big advantage here is that the particulars of the locations of the cameras with respect either to the egg or to each other won't matter. There are a few easily achieved requirements, such as cameras' visual access to the surface of interest (prior to entry into an image of the robot arm), good physical-space-to-camera-space resolution and good angle of camera separation. But by and large the cameras can be located in convenient places. As discussed in the final chapter prospective robot-guiding cameras can even be panned and tilted automatically to be in a position to contribute to the maneuver at hand. Another freedom associated with working in camera space is just as welcomed: An understanding of the globally complex geometric properties of the lens mapping - where in the image photographed points in 3D physical space will fall - isn't needed for realization of the camera-space criteria above to be a safe guarantee of desired physical pen-tip positioning.

Mathematically, the problem comes down to solving for the angles θ_1, θ_2, θ_3 that produce

$$
\begin{aligned}
x_{c1}^t - f_{1x}(\theta_1, \theta_2, \theta_3, \theta_{4o}, \theta_{5o}, \theta_{6o}) &= r_1(\theta_1, \theta_2, \theta_3) = 0 \\
y_{c1}^t - f_{1y}(\theta_1, \theta_2, \theta_3, \theta_{4o}, \theta_{5o}, \theta_{6o}) &= r_2(\theta_1, \theta_2, \theta_3) = 0 \\
x_{c2}^t - f_{2x}(\theta_1, \theta_2, \theta_3, \theta_{4o}, \theta_{5o}, \theta_{6o}) &= r_3(\theta_1, \theta_2, \theta_3) = 0 \\
y_{c2}^t - f_{2y}(\theta_1, \theta_2, \theta_3, \theta_{4o}, \theta_{5o}, \theta_{6o}) &= r_4(\theta_1, \theta_2, \theta_3) = 0
\end{aligned}
\tag{6.1}
$$

which leaves four equations for three unknowns, θ_1, θ_2 and θ_3. This is not a problem theoretically or practically if we know in advance that there should exist a combination of θ_1, θ_2 and θ_3 consistent with satisfaction of all four equations. And that should certainly be the case here. However, the presence of the "redundancy" – the extra relationship – is actually a very good thing for two reasons. First it creates an internal check (think of "bells and whistles" in software - contingency provisions in the event of an anomaly) as to the compatibility of the various data going into the solution for the robot's angles, and hence an indication of the presence/absence of normal operation. And second, the redundancy permits the use of a kind of averaging - the same kind of advantage you have when you sample several times some scalar quantity like temperature and take an average to filter out random variations in each sample.

Consider the averaging effect first. If we were in fact sampling temperatures and had, say, four samples, $T1$, $T2$, $T3$, and $T4$, one approach to finding an average would be to sum the four and then divide by four: $Tavg=[T1+T2+T3+T4]/4$. Another approach would be to find the value of $Tavg$ that minimizes $J(Tavg)=(T1-Tavg)^2+(T2-Tavg)^2+(T3-Tavg)^2+(T4-Tavg)^2$. This is called the method of least squares, and it extends directly to our problem of finding θ_1, θ_2 and θ_3 from the "over-prescribed" conditions of Eqs. 6.1: We solve for the three joint rotations that minimize $J(\theta_1,\theta_2,\theta_3) = r_1^2(\theta_1,\theta_2,\theta_3) + r_2^2(\theta_1,\theta_2,\theta_3) + r_3^2(\theta_1,\theta_2,\theta_3)+ r_4^2(\theta_1,\theta_2,\theta_3)$. Appendix A describes one particular approach and algorithm – one of several possibilities – that could be applied to either of these least-squares problems.

Concerning our example of the use of averaging to improve a temperature estimate, there is an important practical consideration: If the primary reason for error in any individual sample were an out-of-calibration thermometer, then, using only samples from this faulty instrument, the benefit of averaging would be lost. With more and more samples taken into consideration we would approach a certain value of the average, but - least squares or simple mean - the number would be wrong. This observation speaks to the use of Camera-Space Manipulation as opposed to calibration: If the individual camera samples were combined to determine directly the absolute physical coordinates of our laser-spot target point then camera calibration would have to be entailed. No amount of current-maneuver sample averaging would reduce the effects of inevitable error in such calibration. Bypassing calibration and going directly to camera space however – the reference frames of the sensors themselves - eliminates this possibility as well as the tedium of calibration. That's why humans can thread a needle, even though none of us have ever had our eyes calibrated; we construe and pursue all maneuver objectives in the reference frame of our own visual sense. Yet what machines can do that we would find extremely difficult is to combine the perception of several "eyes", several such reference frames scattered all over the room, each with the advantage of its own perspective, into one grand reconciliation of the robot movement that best satisfies all eyes at once.

To improve both the geometric advantage of multiple perspectives and the statistical advantage of reducing random, near-zero-mean error through averaging of redundant information, more than two cameras can therefore be used. This ability to apply multiple, independent cameras to the guidance of a single rigid-body-positioning event will be extended. Certain cameras can be positioned in such a way as to be "responsible" for guiding one end of a very large shipping crate into one side of a tight-fitting receptacle, for example, while other groups of cameras are ensuring simultaneous insertion of the other end.

We do not say that multiple cameras should be used to compensate for the fact that, for any given maneuver, one or more of them may have poor visual access and therefore unreliable resolution of the laser-spot-determined target point(s). In this case the best way to exploit extra cameras is to leave out those with poorly resolved visual access to the target of the current calculation and maneuver. Cameras today with standard lenses are relatively cheap; CSM with its ability to apply multiple cameras, each with its own "demands" on the maneuver outcome, is a good way to take advantage of this low cost.

Resolution, then, of the $\theta_1,\theta_2,\theta_3$ we will command the first three servomechanisms to attain is achieved by minimizing $J(\theta_1,\theta_2,\theta_3) = r_1^2(\theta_1,\theta_2,\theta_3) + r_2^2(\theta_1,\theta_2,\theta_3) + r_3^2(\theta_1,\theta_2,\theta_3)+ r_4^2(\theta_1,\theta_2,\theta_3)$, where the residuals r_1 and r_2, defined in Eqs. 6.1, pertain to Camera 1, and where r_3 and r_4 pertain to Camera 2. As mentioned, provided they have good access to the target points, more of the advantages of averaging can be gained by including residuals for additional cameras that may participate in guiding the current maneuver. Unlike the averaging of temperatures from a single faulty thermometer, the combining of cameras in this way brings in observers that have very little to do with each other until the minimum J is found. Each camera locates the laser spot target, for example, on its own, without reference to the others. And each is involved in the estimation of its own camera-space kinematic relationships almost completely independently of the other(s). (There is one small correlation in error among the various participant cameras: The same faulty nominal kinematic model factors in to estimates of the camera-space kinematics of each; however, as discussed in the next chapter, this effect should be very small.) The near independence of the various cameras' maneuver requirements, combined with the aspect of working directly in camera space thereby avoiding calibration, and exploiting the geometric advantage of each of several points of view, makes the application of multiple-camera redundancy result both in high reliability and high three-dimensional precision.

One advantage, then, of multiple independent cameras is that they afford an automatic internal check on the current maneuver. Regarding the four Equations 6-1, it was stated that there should be no conflict among the four and hence no problem having four equations to solve for the three unknowns $\theta_1, \theta_2, \theta_3$. In reality small, primarily random error, mostly related to the pixel quantization of the camera-space domain, will create some small conflict; hence the benefit of averaging. The value of the minimum J itself is a good indicator of this conflict for any given maneuver. As the number of participant cameras grows so too does this benefit of averaging grow. Particularly with large numbers of participant cameras, the minimized index J provides great assurance of the normalcy and hence expected actual precision of the current maneuver. If this least J is very close to zero then indeed there is no conflict among several near-independent observers (*Figure 6-6*). If on the other hand J departs significantly from zero we have a built-in diagnostic: Is just one camera responsible for this anomaly? In other words, if a particular camera is left out completely from the calculation, does J return to its normal low levels? Or is there something more pervasive going on? Such prospects for self-diagnosis and self-correction are very valuable for the kind of system autonomy we seek.

Figure 6-6. A near-zero number for the minimized J is assurance of agreement among independent cameras regarding the joint angles that will get the job done. Independence of several camera-observers is near-assurance of maneuver success

In Chapter 2 it was discussed that the individual rotational servomechanisms for a robot respond reliably to reference inputs. They are capable of adjusting the voltage input, for example, to each motor as needed to just reach the reference angles. The above discussion pertains to a way of *calculating* these reference values. In the context of the block diagram for a given servomechanism – *Figure 2-5* – the rotations calculated above would be the reference θ_r. (With teach/repeat these values are established by the human teacher.)

Determination of the camera-space kinematics

The other question mentioned at the beginning of this chapter involves how we arrive at the critical expressions for the camera-space kinematics, e.g. $f_{1x}(\theta_1,\theta_2,\theta_3,\theta_4,\theta_5,\theta_6)$ particularly with such unreliable "ingredients" – namely the robot's 3D forward kinematics model which is seldom accurate globally even to within an inch, and the camera/lens characteristics. Three ideas are involved in addressing this point:

1. Very high precision of the functions f_{1x} f_{1y} f_{2x} f_{2y} (if just 2 cameras) is not, as a practical matter, required until the current maneuver is close to its terminus. The reason has to do with the relative speed with which today's computers can minimize $J(\theta_1,\theta_2,\theta_3)$, using any update of the functions f_{1x} f_{1y} f_{2x} f_{2y} . Robot advancement in the direction of the terminus early in a maneuver can be based upon rough initial estimates of the functions f_{1x} f_{1y} f_{2x} f_{2y} , which, as indicated below, can be refined during approach.

2. Stationary cameras that have previously sampled the laser-spot target are usually also in a position to see the tool or end effector as it approaches the target. It is technically straightforward to sample actual internal-joint values, $\theta_1,\theta_2,...\theta_6$, at the same instants when such approach poses are registered in participant-camera image samples. Because image analysis of the end effector is demanding in terms of computer execution time, we are fortunate there is no need for near-instantaneous processing of the images. Unlike visual servoing, CSM does not apply current-maneuver images instantaneously as acquired. Rather, course corrections, much as a navigator might make in steering a ship, allow for accommodation of time delay. In fact, there is a big advantage in controlling a holonomic robot whose position is algebraically related to the control input (the reference angles of the joints) in this way. In the case of a ship, where position is only differentially related to control inputs, such as rudder angle or engine speed, drift occurs following each course correction. The robot's *algebraic* predictions of the future input-output relationship allow for the operation to be executed open-loop - that is without ongoing monitoring of progress with respect to the targeted terminus.

As long as the input (the needed joint rotations at the terminus) can be guaranteed, which we know from Chapter 2 today's rotational servomechanisms do guarantee, the output – delivery of the tip in camera space – will follow. The rapid, "confident," and direct delivery of position of teach/repeat robots in a factory is due to this excellent property. Of course much effort and expense will have gone into ensuring that the confident open-loop motion of teach/repeat is met by a previously, precisely prepositioned workpiece, or else a collision is likely. With CSM similar robot-motion confidence is warranted by the much cheaper fact that cameras have registered the actual location of the workpiece. Such efficient, assertive motion is something humans seldom use. The only time we deliver a load with this level of assertion is when we intend to strike, as in a boxing round for instance. The car-key-positioning exercise of Chapter 2, for instance, involves an ongoing referral to the external key-vs-pencil relationship, yielding a more tentative kind of control, and a need for visual access all the way up to the terminus. The position servomechanism on the other hand only compares its current internal angle(s) of rotation to the reference; and this joint-angle feedback reading is essentially instantaneous and perfect.

The course corrections that are required by CSM en route to the terminus are not in fact due to the kind of drift to which systems with differential input-output relationships are prone. They occur because of actual updating, small fine-tuning, of the camera-space-kinematics functions, f_{1x} f_{1y} f_{2x} f_{2y}. As mentioned, although, even with today's fast computers, the updating process takes a finite amount of time, it can occur even as the physical maneuver proceeds toward its terminus. Trajectory modification can be applied if/as new visual information has been processed prior to maneuver termination.

3. The reason the kinematic mapping of six-dimensional "joint space" ($\underline{\boldsymbol{\theta}}=[\theta_1, \theta_2, \ldots\theta_6]^T$) into the two-dimensional Camera-1-space location of our white X, $f_{1x}(\underline{\boldsymbol{\theta}})$, $f_{1y}(\underline{\boldsymbol{\theta}})$, must be updated, albeit only by very small increments and only a couple of times per approach, is that the model bases for these functions are not globally precise.

The words "mapping" and "global" call to mind something like *Figure 6-7*. Use of the word "map" in this context connotes correspondence. In the representation of *Figure 6-7* it would be a one-to-one correspondence between any given point on the surface of the physical world and a juncture on the flat representation of the figure. In the case of our kinematics functions $f_{1x}(\underline{\boldsymbol{\theta}})$, $f_{1y}(\underline{\boldsymbol{\theta}})$ the map is a many-to-one correspondence between the six joint angles and location of the pen tip's white X in Camera-1 space.

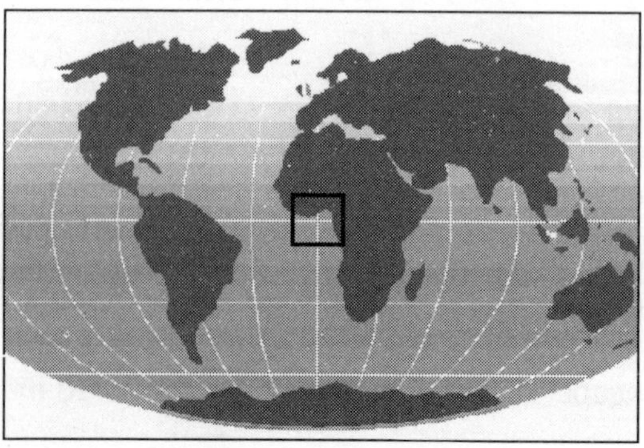

Figure 6-7. The algorithm used to create a one-to-one mapping between points on the physical world and this flat representation is neither mathematically simple nor intuitive – except, asymptotically, in a region near the map's center, on the equator near the western coast of Africa.

Consider the region maybe a thousand miles square indicated in the center of the map. Within this region, a mathematically simple and intuitive correspondence between points on the map and points on the physical surface of the earth works well: Up on the map means north; right on the map means East; and any given distance between two points on the map is proportional to the straight-line distance of travel across the face of the earth. With these simple rules in mind, identifying the locations of two known spots, say two city centers, with junctures on the map is all the calibration needed. From there, the map's location of a third or fourth city center would be enough to calculate the exact distance from that city to any of the others, along with direction. Very straightforward.

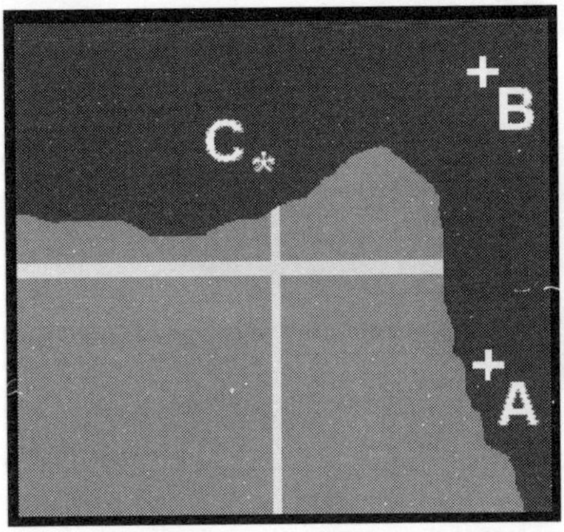

Figure 6-8. Any thousand-mile-square patch of the world could be treated with the same simple map formulation: Up is North; right is East; and the straight-line distance between any two points on the surface of the earth is proportional to the distance separating them on the map.

This simplicity breaks down as you get further and further from the contained region, however. Suppose that you wanted to include the North Pole on your flat map. According to the above rules it would be proportional to $\pi R/2$ straight up from the map center, where R is the radius of the earth. Similarly a point a quarter of the way west around the world would be $\pi R/2$ to the left. We know that physically the distance between this new point and the North Pole should be $\pi R/2$ again, but your map would put it at $\pi R/2^{1/2}$, almost fifty percent more than actual. In fact sticking with such mapping of these distant points would require an accommodation and distortion of all other points that would make most of the continents unrecognizable. The mathematically complex algorithm used to create *Figure 6-7* results in similar distortions, although the straight-line-distance error it creates in terms of the above test is less dramatic, and the resulting shapes of most of the continents leaves them recognizable.

Now all of that assumes that the world is actually a perfect sphere. In fact, it departs slightly from that, bulging out here and there. But the simple mapping model continues to work well in our *limited* region even in the presence of these "real-world" bulges. Applicability of the asymptotic limit continues to hold in our contained region regardless of these larger-scale departures of the exact whole from a perfect sphere. Moreover, we are not really limited in our use of this asymptotic limit in the following additional sense: We could pick up our black square of *Figure 6-7* from the equator on the western coast of Africa and plunk it down anywhere in the world we wanted. No real need to remain true to the conventional definitions of north and east, in fact, so long as we're consistent in our mapping, and these directions are orthogonal and right-handed ("east" is to your right if you're facing "north").

Figure 6-9. While seemingly very different, these two modes of transportation – SUV and GPP (gigantic pen plotter) – have something in common: Internal mechanical rotation is what ultimately transports our user from B to C.

One way to proceed creating the one-to-one correspondence or "mapping" between points within the thousand-mile-by-thousand-mile region is to identify two points, say the center of City A and the center of City B, as indicated in *Figure 6-8*. Call "north" the direction of the line on your map that goes from A to B. Then "east" would be perpendicular to this direction, and to the right. In keeping with the rules of our simple mathematical formulation, any other point on the map, say the center of City C, would be a distance from A or B that is proportional to the corresponding physical distances across the earth's surface (leaving aside hills, valleys, etc.) Because our asymptotic limit to the near-spherical-but-bulgy earth is applicable across any suitably small region of the earth's surface, there will be consistency in direction between the map's rendering of C's location relative to B, or C's location relative to A. In fact, all physical-surface points can be located in a similar, consistent way onto this map. Using the same simple rules, every city center and landmark in the region is consistently represented with respect to every other. Points along the coastline, for example, can be located with respect to previously mapped references, and thereby placed onto the map. Enough of these can be mapped to provide an accurate picture of the shape of the coast. To a user of the map the important attribute is its predictive ability. If you were located at B and wanted to go to C, the map would tell you how far to travel and in which direction.

An interesting and important element of CSM can be illustrated by considering the actual mode of transportation that our map user might take in traveling to City C from B. In particular, consider the two alternative conveyances pictured in *Figure 6-9*: SUV vs. gigantic pen plotter (GPP).

While very different in most respects, the two modes of transportation have one attribute, although not an obvious one, in common: Both succeed in transporting our map user by way of motor-driven mechanical rotation. In the case of the SUV the motor rotates the wheels of course. And in the case of the pen plotter, although they are not shown, two rotational motors are actually responsible for movement of the pen's x-coordinate and y-coordinate, one motor for each. A rotational sensor, an optical encoder, measures these GPP rotations even as it serves as part of the rotational servomechanism that enforces reference trajectories for the x and y components of position. A special mechanism converts the motors' rotational motion to the pen's translational motion.

If the x-y pen plotter returns each of its two optical encoders to the same readings they had when it last visited City C, it will not matter where the device traveled in the meantime, or what disturbances it encountered: the pen tip will return to City C. This property of the device with internal encoders is "holonomy". Note that neither the SUV nor, for that matter, the car-key positioner of Chapter 2 can say the same thing. While wheel rotation of the SUV is responsible for its ultimate ability to transport, and while finger-, wrist- and elbow-joint rotation of the key pusher is responsible for getting the key to its goal, both of these systems are nonholonomic. As discussed in detail in Chapter 9 this means, among other things, that there is a path dependency on the relationship between internal rotation and external position of the object of interest. As a practical matter, this means that one must continually, with a nonholonomic system, refer control to external considerations: Drivers must watch the road and key pushers must keep an eye on their key. Also with nonholomomic systems, disturbances are not generally rejected. Disturbances will influence the trajectory and the future relationship between internal (joint or wheel) rotation and position of the body of interest.

With our gigantic pen plotter, however, it is a different - and advantageous - story. For example, every time it passes through City A during an A-team home game it encounters a lot of resistance while drawing its mark through the stadium parking lot. But it goes on, correcting itself and, importantly, sustaining that fixed relationship between internal encoder readings and pen-tip position. In fact, if the map reader knew in advance that he would always travel by way of x-y pen plotter, he may have reason to calculate a different kind of map. A more useful map could be one that gives geographic position as a function of θ_1 and θ_2, the two internal-angle positions of the device. Because once these two angles have been specified, the pen-tip transporter need not refer to external progress at all. In this respect its motion is open-loop: "Just servo me to City C's internal coordinates, (θ_{1r}, θ_{2r})." Of course altering the map to relate directly GPP internal coordinates to physical pen-tip location would add a small amount of complexity: *Figure 6-9*'s angular difference between pen-arm orientation and "east" or "north" would have to be accounted for, for instance. But provided the region of land over which the map is defined remains small, it is easy to find a mathematical form for the map that is understood, simple and locally precise.

A similar insight enables estimation of the camera-space kinematics using a simple model, an asymptotic limit of the global reality. Camera-space kinematics can be based upon local, simultaneous samples of joint rotations and pen tip appearing in camera space, and it becomes extremely accurate locally. But the global models upon which the local model is based differ from the more complicated and subtle reality of optics and kinematics - a reality much like the true contour of our bulging earth, which is not simple and perhaps gradually shifting and not knowable.

The local goodness of the simpler model extends in another way: As indicated in *Figure 6-10*, it is not practical to sample the pen tip in camera space. However, other surface points that move as a rigid body with the tip both can be seen at least intermittently and are not normally affected by shadows of end-of-maneuver closure. Moreover, these have precisely known geometric location on the end member with respect to the tip point of interest. They are also fairly close to the pen tip. As discussed in the next chapter, use of the same asymptotic limits that allow for the application of approach points to determine camera-space kinematics applicable to the maneuver terminus also permits the proxy use of these visible junctures in the quest to refine to a high degree of predictive precision the camera-space kinematics that apply to our positioned point of interest – the invisible white X embedded in the pen tip of *Figure 6-5*.

Orientation control

It was mentioned earlier that the camera-space kinematics used to position the white X can also be used to control the wrist angles, θ_4, θ_5, θ_6, with the goal of achieving simultaneous orientation control. This procedure necessarily entails consideration of the direction that is perpendicular to the egg shell at the point of interest. Such information is not available from the single laser spot converged onto the spot center. However, it can be inferred in a way that is consistent with CSM-based robot control by taking advantage of the asymptotic limit involving curvature of continuous surfaces.

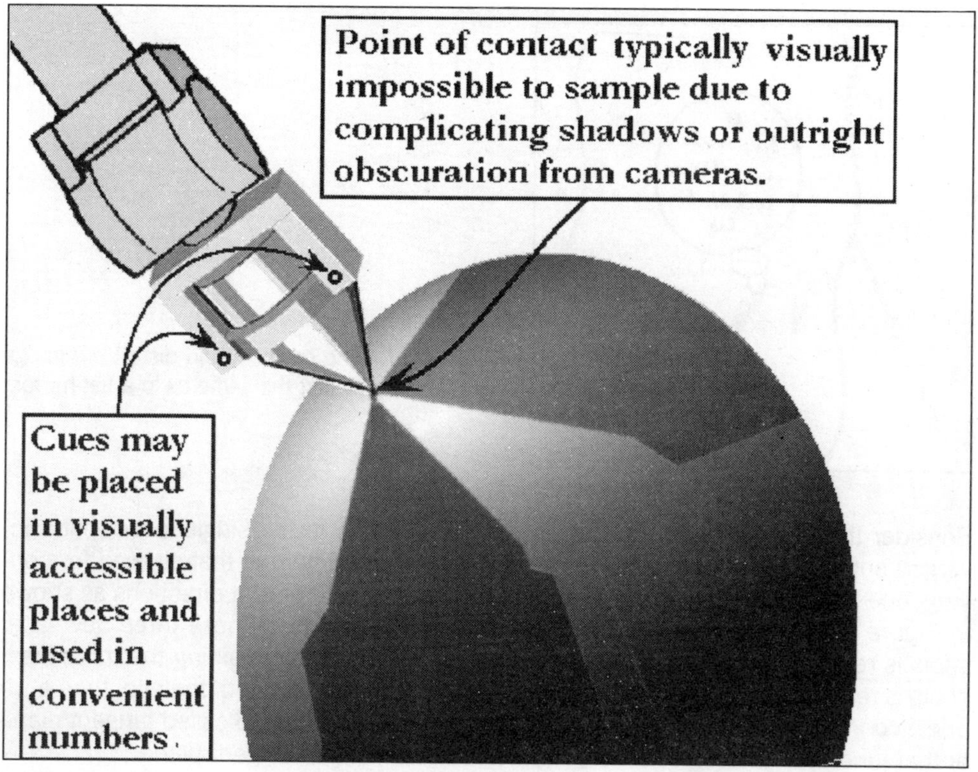

Figure 6-10. Use of an appropriate asymtotic limit permits application of visually more accessible points in the estimation and refinement of camera-space kinematics for our point of interest, the pen tip.

Think of your image in a mirror. If the mirror is flat, then what you see is a fair representation of what others would see. Consider next a distorted mirror, the kind you might see at an amusement park. The "departure from flatness" means that your reflection is discernibly different from reality.

But consider the idea of "scale". Think of a fly looking into the same mirror, as shown in *Figure 6-11*. The region of the mirror in which the fly is reflected is small relative to the scale of the mirror's distortion. And the effect of this is that the fly's image is no different from what would appear in a flat mirror. Similarly, as noted earlier, the scale within which we ordinarily travel is small compared to the curvature of the earth. So the earth seems flat.

Figure 6-11. The mirror would distort *our* image but the fly looks just the same as in a flat mirror.

Consider the egg curvature. Suppose that, in addition to the single laser spot converged onto the supervisor-selected point, the same pan/tilt unit that allowed for convergence to that point displaces the spot slightly in three different directions as shown in *Figure 6-12*. As with the originally selected point, each of these three new laser spots is registered in all participant cameras. If the region containing the three laser spots is reasonably small compared with the radius of curvature of the egg, then a triangle connecting the three in physical space would define the so-called tangent plane at that juncture. The unit vector **n** is perpendicular to this tangent plane.

Referring to *Figure 6-13*, note that for all three of these new points A, B and C, each identified in all participating camera spaces using image differencing as per the original point, a minimization of $J(\theta_1,\theta_2,\theta_3)$ can be accomplished just as if the goal were to locate our pen tip at each of them individually. Without actually commanding the robot to move to any of them, we can compute the first three joint angles $\theta_1,\theta_2,\theta_3$ that correspond with each of junctures A, B and C. As indicated in *Figure 6-14*, we now return to the forward kinematics model for this robot, albeit a globally errant one. We use the previously asserted $\theta_{4o},\theta_{5o},\theta_{6o}$ together with each of the three points' $\theta_1,\theta_2,\theta_3$ to calculate nominal values for the three-dimensional coordinates of A, B. and C. These, in accordance with *Figures. 6-13* and *6-14*, allow for computation of the unit normal **n**.

It should be noted that this **n** is referred to whatever stationary frame of reference has been used to describe the forward, three-dimensional kinematics of the robot. And to the extent that the global forward kinematics are wrong (a considerable extent we assert), an understanding of **n** in conjunction with that frame would be similarly in error. Fortunately, the only use we make of **n** is as a reference direction for finding joint angles that will point correctly the pen on the end effector. This pointing calcu-

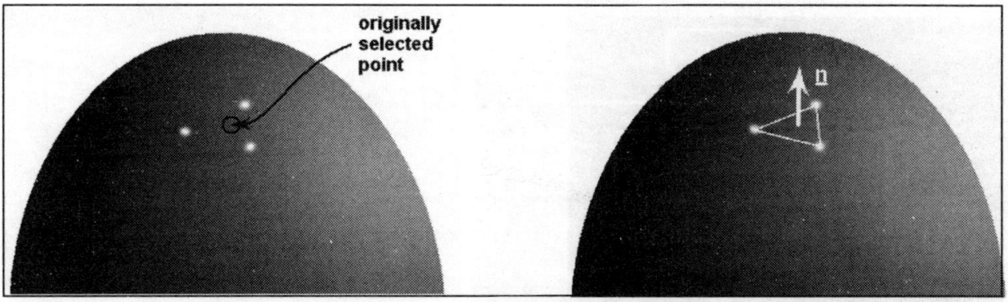

Figure 6-12. The precise location of the three laser spots, provided they don't come close to forming a straight line and provided they are all near to the originally selected point, will not matter much to the finding of the tangent plane with its unit normal **n** at our point of interest.

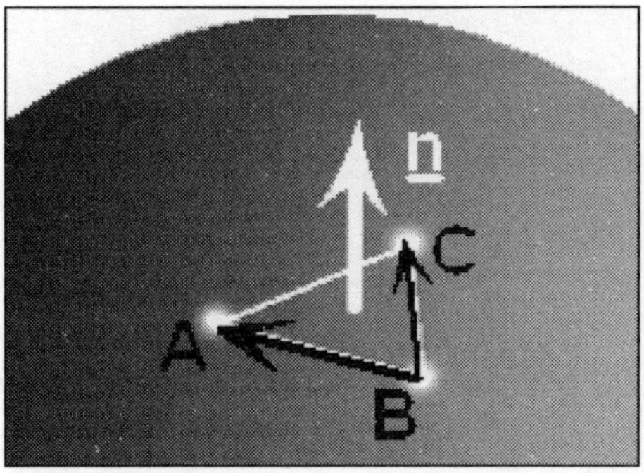

Figure 6-13. If points A, B and C are all close to our originally selected point, then the unit vector **n** will be in the direction of the cross product of the vector **BC** with **BA**.

lation in turn takes advantage of the same errant nominal-forward-kinematics model used to establish **n**. The errors then, which are characteristic of that particular local region of joint space, cancel – with the result that the end-member's unit normal e identically opposes **n**. The need for iteration in the flow chart of *Figure 6-14* is a consequence of the full coupling between all six joint rotations, not just the fourth and fifth, and the unit vector **e**, as indicated in *Figure 6-15*. Nevertheless, for any fixed value of the other angles, only two angles – we choose θ_4 and θ_5 - are needed for purposes of achieving a dot product of **e** with **n** equal to –1. This is done with the use of the nominal forward kinematics model as discussed in Appendices A and B. The last angle θ_6 is held constant throughout the maneuver. In fact, depending upon how the pen is aligned with the final axis of rotation of the robot, θ_6 may not actually affect the pointing of the pen tip at all. If it were important to rotate the tool to a specific angle about its axis, such as the case for example of a pen knife making an incision where the blade must be aligned with the direction of the cut, this final axis would come into play and a similar kind of iterative procedure would be involved in finding all six angles.

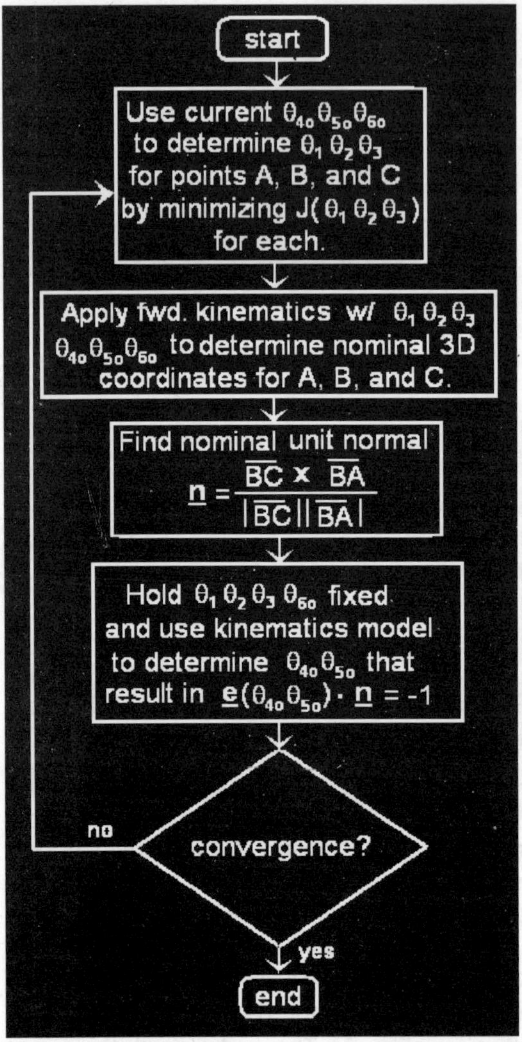

Figure 6-14. Only two of three wrist angles are needed to point the pen in the right direction. These are calculated iteratively using information from three laser spots directed onto the egg near the originally selected point.

Importantly, in the flow chart of *Figure 6-14*, the first three angles $\theta_1, \theta_2, \theta_3$ that are "held fixed" during the procedure for correcting orientation by recomputing θ_{40}, θ_{50} are found as described earlier in this chapter, using the previously determined θ_{40}, θ_{50} – together with the fixed θ_{60} – as before in order to collocate the white X with the originally selected target. An important prospect here, however, is that useful control of the tool's approach is, at this iterative stage, very straightforward. As indicated in *Figure 6-16* and detailed in Chapters 7 and 8, it is quite easy to apply all accumulated camera-space data of the end effector to the estimation of camera-space kine-

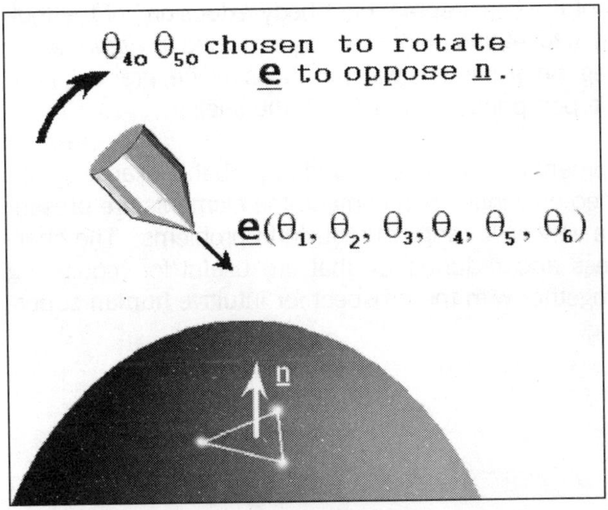

Figure 6-15. Pointing correction.

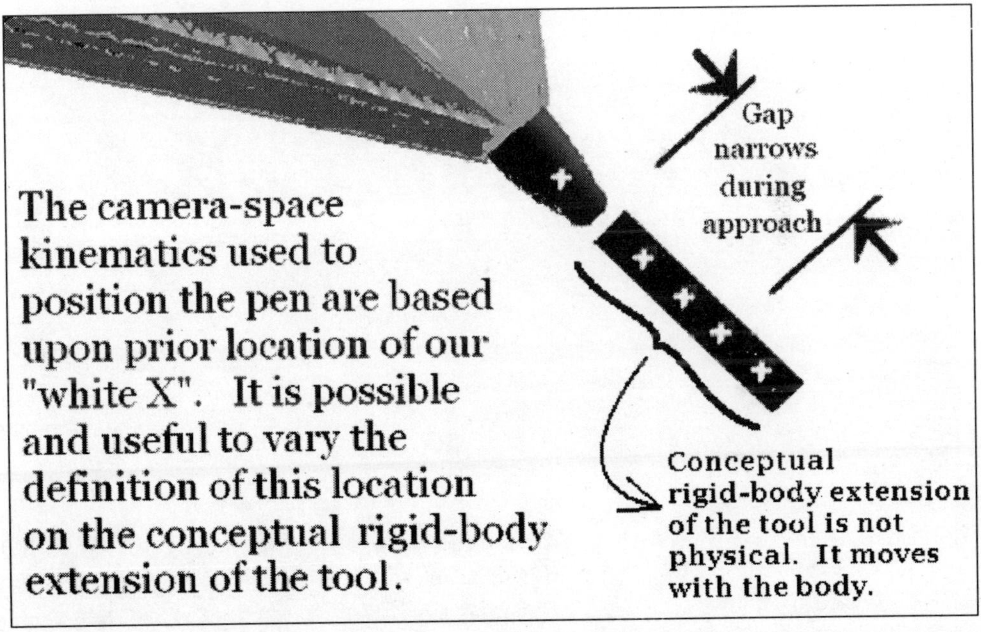

Figure 6-16. With the iterative computation of both tip position and pen orientation it is possible to advance the pen gradually and deliberately toward the egg along its own writing axis by sequentially redefining the location of the white X.

matics associated with any point of the "conceptual rigid-body extension" of the tool. A natural choice for our approach is to let this point slide, as indicated in *Figure 6-16*, along the pen's axis. Commanding the arm in this way produces a nice, gradual, controlled-orientation approach of the pen prior to contact with the shell.

Although the present development uses a peculiar and improbable example, and even for it we have so far discussed only initial pen contact, the elements are present to apply mechanical dexterity to a very wide range of real-world problems. The characteristics of precision, robustness and redundancy that are useful for monitoring maneuver quality automatically together with the prospect for intuitive human supervision are present across this range.

CHAPTER 7

CAMERA-SPACE KINEMATICS

If Chapter 6's premise is true - that manipulation tasks are realized by positioning selected end-member junctures within the 2D spaces of at least two well-separated cameras - it comes down to this: In real-world circumstances can we apply image samples and simultaneous joint-rotation samples, acquired en route to the terminus, to determine, with enough precision and timeliness, the "camera-space kinematics"? So important are these algebraic relationships between a robot's internal joint angles and location of the positioned point in camera space that nothing is taken for granted: The present chapter emphasizes direct verification of estimates at every step of development.

The concept of a "rigid body" is an important one for us here. It is the idea that an object's individual, constituent parts sustain a fixed geometric relationship one to another, regardless of how the body overall may move. If you could embed a Cartesian coordinate system within the body, such as Chapter 6's pen-plot tool shown here in *Figure 7-1*, then the coordinates of every juncture within a rigid body never change. In classical mechanics the "rigid body" is contrasted with the "deformable body". This distinction, as much as it may have to do with their material or connectivity, entails the size of forces to which each type may be subjected, and how small any deformation must be to be considered negligible (all objects deform some when acted upon by forces). For purposes of the present discussion we consider the end effector to be a rigid body. In particular, the geometric relationship between the surface-cue centers of *Figure 7-1* and the positioned point on the pen tip is permanent.

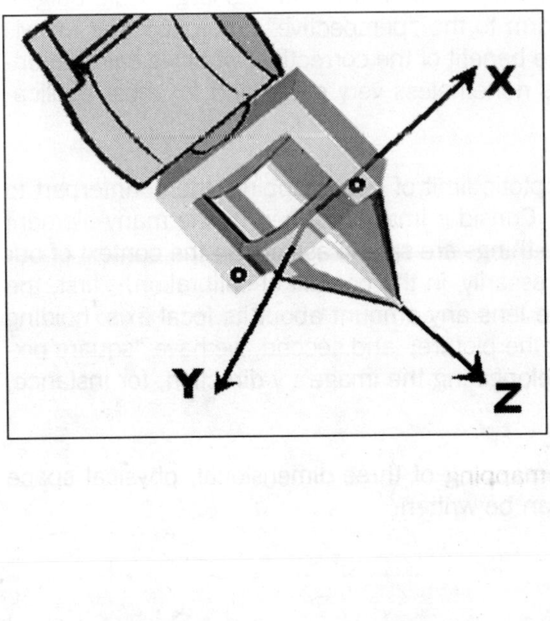

Figure 7-1. If coordinate axes could be embedded into a rigid body, then any of the body's constituent pieces would sustain the same coordinates relative to these axes, regardless of body motion.

Not only is this relationship between cues and positioned point(s) permanent, it is also precisely known, *apriori*. That translates into known coordinates of cue centers with respect to the reference frame of *Figure 7-1* as well as known coordinates of any positioned points on the conceptual rigid-body extension relative to the same frame. Inaccuracy here will produce similar inaccuracy of physical positioning. This requirement typically is not too demanding in practice, much less demanding, for example, than building an entire robot with actual kinematics similarly close to their nominal model.

Nevertheless, the nominal kinematics are employed in the process of estimating camera-space kinematics locally. It's just that the kinematic error's effect on our estimates becomes negligible as we apply in effect an asymptotic limit – albeit a not-quite-perfect one – of these *apriori* geometric relationships to our estimation model. The other half of the estimation model is a perfect asymptotic limit of actual lens mapping of points in the three-dimensional world into camera space. But although mathematically perfect the size of the piece of physical space over which this limit is adequate is for many applications just too small. So we expand it with an improvement. This improvement has an analogy with the extent of terrain over which the "flat-world" limit can be applied accurately in map-making, as discussed in Chapter 6. As the boundaries of the map extend further and further, the flat-earth rule – physical distance between two points on a flat map is proportional to their over-land separation - breaks down. But it would last longer, we could extend the range further, if we drew the map on a surface formed like a part of a sphere. Even if the radius of curvature of this sphere were a bit too large or a bit too small in proportion to the earth's radius, just coming close would increase the accuracy of the map greatly. That remains true even though the actual earth isn't quite a sphere at all, but rather has large-scale bulges. Even so, any real lens does not conform to the "perspective" correction that we will apply to our estimation model. But the benefit of the correction, what we call "flattening", in producing useful estimates is nonetheless very great, and for most applications, enabling.

We begin with the simplest asymptotic limit of lens mapping, the counterpart to the flat-earth model of real mapping. Consider images formed by the many-element lens of the camera of *Figure 7-2*. Two things are safe to assume in the context of our estimation-model building (if not, necessarily, in the context of calibration): first, the lens is radially symmetric; rotating the lens any amount about its focal axis, holding the image plane fixed, would not alter the picture; and second, we have "square pixels": there is no electronic distortion elongating the image's y-direction, for instance, relative to its x-direction.

The general form of continuous mapping of three-dimensional, physical space into two-dimensional camera space can be written:

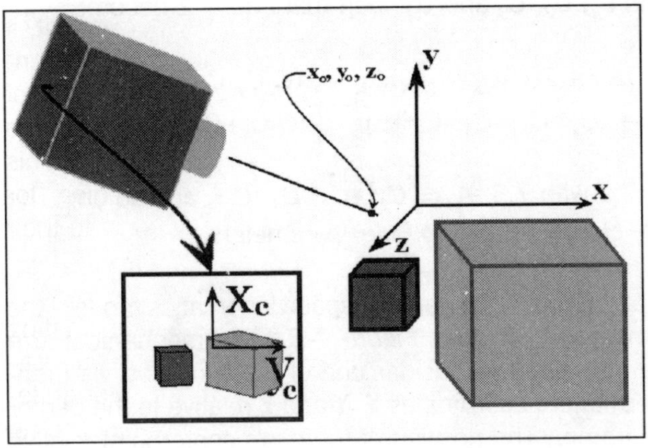

Figure 7-2. Any point x, y, z that is in focus and in view of our camera will have a "mapping", an actual position or pair of coordinates in camera space. The actual relationship here depends upon the lens and electronics, and is complex, almost impossible to determine globally.

$$x_c = f_x(x,y,z)$$
$$y_c = f_y(x,y,z) \tag{7.1}$$

where x_c and y_c are the coordinates, as indicated in *Figure 7-2*, of a photographed juncture in camera space, and where x, y, and z are coordinates of the point in three-dimensional physical space. Consider next any physical-space point x_0,y_0,z_0 that happens to lie along the camera's focal axis, as indicated in *Figure 7-2*. Assuming that the origin of camera space is defined to lie in the middle of the image, the camera-space coordinates of this point are, in accordance with Eqs. 7.1, $x_c = f_x(x_0,y_0,z_0)=0$, $y_c=f_y(x_0,y_0,z_0)=0$. Next, consider a small but arbitrary departure, in physical space, from x_0,y_0,z_0. That is, consider $x=x_0+\Delta x$, $y=y_0+\Delta y$, $z=z_0+\Delta z$. A continuous lens mapping means that, for sufficiently small $\Delta x\ \Delta y\ \Delta z$,

$$x_c = A_{11}\Delta x + A_{12}\Delta y + A_{13}\Delta z$$
$$y_c = A_{21}\Delta x + A_{22}\Delta y + A_{23}\Delta z \tag{7.2}$$

where A_{11}, for example, would be the partial derivative of the unknown function $f_x(x,y,z)$ with respect to x, evaluated at the point x_0,y_0,z_0.

Given the assumptions mentioned earlier - radial symmetry of the lens and no bias or distortion between the camera's horizontal and vertical direction - there are two consequent constraints among the six elements A_{ij} of Eqs. 7.2. In particular, it can be shown that $A_{11}^2 + A_{12}^2 + A_{13}^2 = A_{21}^2 + A_{22}^2 + A_{23}^2$. Also, $A_{11}A_{21} + A_{12}A_{22} + A_{13}A_{23}=0$. These two constraints are imposed automatically by replacing the six original elements with four constants C_1, C_2, C_3 and C_4 such that Eqs. 7.2 become

$$x_c = (C_1^2+C_2^2-C_3^2-C_4^2)\Delta x + 2(C_2C_3+C_1C_4)\Delta y + 2(C_2C_4-C_1C_3)\Delta z$$

(7.3)
$$y_c = 2(C_2C_3-C_1C_4)\Delta x + (C_1^2-C_2^2+C_3^2-C_4^2)\Delta y + 2(C_3C_4+C_1C_2)\Delta z$$

In other words, comparing Eqs. 7.2 with 7.3, $A_{11}= C_1^2+C_2^2-C_3^2-C_4^2$, and so on. The four elements C_1, C_2, C_3, C_4, are closely related to Euler parameters.

Turning our attention to the other half of the camera-space kinematics model - the robot-mechanism forward kinematics – consider *Figure 7-3*. The mechanical arm enters the image of our camera of interest. Consider one of the two cues, as mentioned earlier with known and permanent coordinates **X**, **Y** and **Z** relative to the pen or other tool's body-fixed reference frame. The position of the cue's centerpoint relative to the fixed x-y-z reference frame is a function of the robot's joint rotations, θ_1, θ_2, ... , θ_n, for an "n" degree-of-freedom robot. On this day, bearing its current load, with the current thermal expansion of the robot's members, true kinematics for the end-member point **X**, **Y**, **Z** and robot of interest are not precisely known. We represent these true forward kinematics with the general model:

(7.4)
$$x = g_x(\theta_1, \theta_2, \ldots, \theta_n; \mathbf{X}, \mathbf{Y}, \mathbf{Z})$$
$$y = g_y(\theta_1, \theta_2, \ldots, \theta_n; \mathbf{X}, \mathbf{Y}, \mathbf{Z})$$
$$z = g_z(\theta_1, \theta_2, \ldots, \theta_n; \mathbf{X}, \mathbf{Y}, \mathbf{Z})$$

where x, y, and z are the coordinates of the one cue measured with respect to the fixed-to-the-world coordinate frame indicated in *Figures 7-2* and *7-3*.

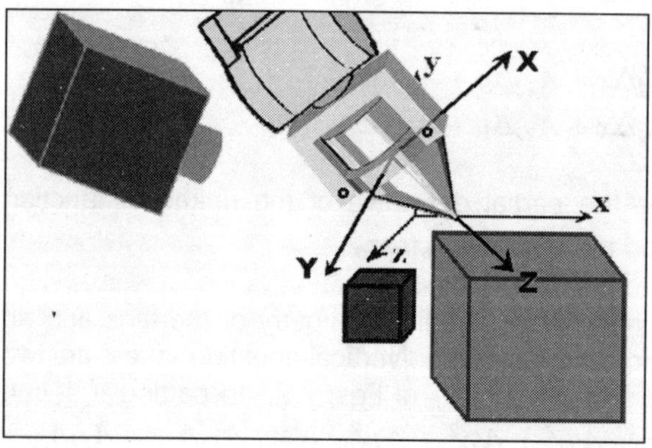

Figure 7-3. As a cue enters the camera's field of view its x,y,z coordinates move as a function of the robot's internal joint angles. Its **X Y Z** coordinates are fixed.

Consider a particular "pose A", θ_{1A}, θ_{2A}, ... , θ_{nA}, say one that we have reason to think is near – maybe just before reaching - the terminus or terminal pose of interest. Let $x_A=g_x(\theta_{1A}, \theta_{2A}, \ldots, \theta_{nA}; \mathbf{X}, \mathbf{Y}, \mathbf{Z})$, $y_A=g_y(\theta_{1A}, \theta_{2A}, \ldots, \theta_{nA}; \mathbf{X}, \mathbf{Y}, \mathbf{Z})$, $z_A=g_z(\theta_{1A}, \theta_{2A}, \ldots, \theta_{nA}; \mathbf{X}, \mathbf{Y}, \mathbf{Z})$ represent the coordinates of the cue in Pose A. Increments in the cue center's position from "pose A", provided corresponding joint changes $\Delta\theta_1$, $\Delta\theta_2$, ... , $\Delta\theta_n$ are small, can be written:

$$\begin{aligned} \Delta x_A &= b_{11}\Delta\theta_1 + b_{12}\Delta\theta_2 + \ldots + b_{1n}\Delta\theta_n \\ \Delta y_A &= b_{21}\Delta\theta_1 + b_{22}\Delta\theta_2 + \ldots + b_{2n}\Delta\theta_n \\ \Delta z_A &= b_{31}\Delta\theta_1 + b_{32}\Delta\theta_2 + \ldots + b_{3n}\Delta\theta_n \end{aligned} \tag{7.5}$$

where b_{11} for example would be the partial derivative of the unknown function $g_x(\theta_1, \theta_2, \ldots, \theta_n; \mathbf{X}, \mathbf{Y}, \mathbf{Z})$ with respect to θ_1, evaluated at θ_{1A}, θ_{2A}, ... , θ_{nA}. These coefficients b_{ij} cannot be determined apriori since the functions g_x g_y g_z are themselves unknown. However, they can be approximated closely using the nominal forward kinematics model, as indicated below.

We haven't said anything yet about the x-y-z coordinate system other than that it is fixed, and hence our cue moves relative to it as the robot joint angles change. We now assume the x-y-z coordinate system to be coincident with that particular frame from which end-member position would be measured if the nominal kinematics perfectly described the end-effector pose at θ_{1A}, θ_{2A}, ... , θ_{nA}. In other words, imagine that a robot was built that perfectly matched the nominal forward kinematic model. This imaginary robot takes on the n angles of pose A, and in doing so the pen-tip tool is located precisely with position and orientation physically identical to those of the same tool where the actual, kinematically imperfect robot is in pose A. The x-y-z frame is defined to be the particular physical frame from which the ideal robot's kinematics would be measured.

Let these known, nominal kinematics be given by:

$$\begin{aligned} x &= g_x^*(\theta_1, \theta_2, \ldots, \theta_n; \mathbf{X}, \mathbf{Y}, \mathbf{Z}) \\ y &= g_y^*(\theta_1, \theta_2, \ldots, \theta_n; \mathbf{X}, \mathbf{Y}, \mathbf{Z}) \\ z &= g_z^*(\theta_1, \theta_2, \ldots, \theta_n; \mathbf{X}, \mathbf{Y}, \mathbf{Z}) \end{aligned} \tag{7.6}$$

Consider next a small rotation from pose A in the last (n^{th}) joint, $\Delta\theta_n$. If (as is reasonably assumed in most cases) the "grasp", the geometry of the tool relative to this last or n^{th} rotation, is very accurately characterized with the nominal forward kinematics model, then $b_{1n}^*=b_{1n}$, and

(7.7)
$$\Delta x_A = b_{1n}{}^*\Delta\theta_n$$
$$\Delta y_A = b_{2n}{}^*\Delta\theta_n$$
$$\Delta z_A = b_{3n}{}^*\Delta\theta_n$$

where $b_{1n}{}^*$, for instance, is the known partial derivative of $g_x{}^*(\theta_1, \theta_2, \ldots, \theta_n; \mathbf{X}, \mathbf{Y}, \mathbf{Z})$ with respect to θ_n evaluated at $\theta_{1A}, \theta_{2A}, \ldots, \theta_{nA}$. As the tool rotates about this furthest-out, n[th] axis, every point on the tool displaces from its pose-A location as predicted by the nominal model. This is factually reasonable, and the general experience in robotics, even if globally the kinematic errors accumulate to inches.

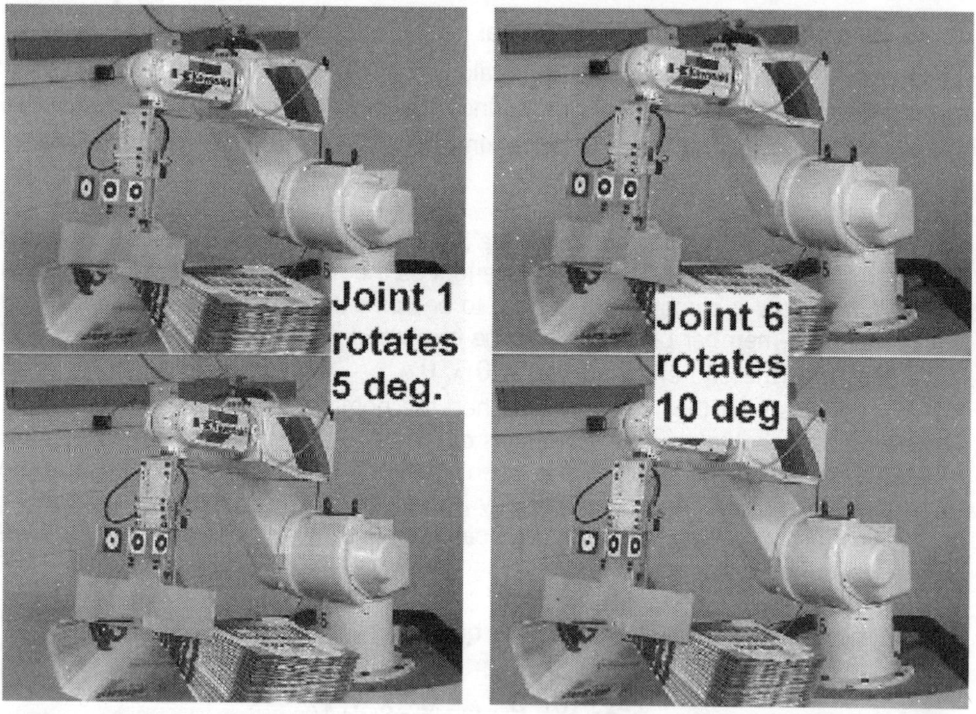

Figure 7-4a. It seems reasonable, intuitively, that the position of each point on the rigid body once-rotated about the sixth and nth axis occupies a location relative to the body's position prior to this rotation as predicted by the differential of the nominal kinematics. What is less obvious is that the same is very nearly true even for the first axis of rotation. Whereas the attendant requirement for the sixth axis is merely correct specification of the geometry of the grasp, the requirement for the first axis also includes good characterization of the angular offsets of the second through sixth joints. Otherwise, however, normal – even large – errors of *global* kinematics characterizations do not affect this result significantly. (Note that the same cannot be said for the nonrigid paper load in the grasp.)

It becomes less obvious, however, as we consider joint axes further and further from the end effector. The counterpart to Eqs. 7.7 in fact remains true by and large, provided all joints' angular offsets are well-calibrated (which type of calibration most robot manufacturers make easy to accomplish). If these are set to within about a quarter of a degree or less, and if the geometry of the grasp of the tool is similarly precisely characterized, there is very little error in the statements $b_{ij}^*=b_{ij}$, $j=1,2,\ldots,n$, and $i=1,2,3$.

If the nominal kinematics perfectly matched the actual kinematics, then as the pose – pose A – about which we find these derivatives is repositioned to a substantially different place in the vast six-dimensional workspace of the robot, the location of the x-y-z frame would remain fixed. However, globally, the nominal kinematics are not terribly accurate. Thus as "pose A" changes substantially so too does the position and orientation of our x-y-z frame shift.

Nevertheless, with the unknown b_{ij} of Eqs. 7.5 replaced by the known b_{ij}^* we have a near-perfect asymptotic limit of the true kinematics near point A. (Yet it is not an absolutely perfect limit because, as each $\Delta\theta_i$ approaches zero the corresponding $\Delta x_A/\Delta\theta_i$, for example, does not become exact.)

In the spirit of drawing a map onto a portion of a spherical surface to extend the region of validity of an asymptotic limit we note that the approximate, but now known, form of Eqs. 7.5, with b_{ij}^* replacing b_{ij}, can be expanded or extended according to

$$
\begin{aligned}
\Delta x_A &= g_x^*(\theta_{1A}+\Delta\theta_{1A},\ \theta_{2A}+\Delta\theta_{2A},\ \ldots,\ \theta_{nA}+\Delta\theta_{nA};\ \mathbf{X},\mathbf{Y},\mathbf{Z}) - x_A\\
\Delta y_A &= g_y^*(\theta_{1A}+\Delta\theta_{1A},\ \theta_{2A}+\Delta\theta_{2A},\ \ldots,\ \theta_{nA}+\Delta\theta_{nA};\ \mathbf{X},\mathbf{Y},\mathbf{Z}) - y_A \qquad (7.8)\\
\Delta z_A &= g_z^*(\theta_{1A}+\Delta\theta_{1A},\ \theta_{2A}+\Delta\theta_{2A},\ \ldots,\ \theta_{nA}+\Delta\theta_{nA};\ \mathbf{X},\mathbf{Y},\mathbf{Z}) - z_A
\end{aligned}
$$

or

$$
\begin{aligned}
x_0+\Delta x &= \Delta x_A + x_A = g_x^*(\theta_1,\theta_2,\ldots,\theta_n;\ \mathbf{X},\mathbf{Y},\mathbf{Z})\\
y_0+\Delta y &= \Delta y_A + y_A = g_y^*(\theta_1,\theta_2,\ldots,\theta_n;\ \mathbf{X},\mathbf{Y},\mathbf{Z}) \qquad (7.9)\\
z_0+\Delta z &= \Delta z_A + z_A = g_z^*(\theta_1,\theta_2,\ldots,\theta_n;\ \mathbf{X},\mathbf{Y},\mathbf{Z})
\end{aligned}
$$

Solving Eqs. 7.9 for Δx Δy Δz and substituting this result back into Eqs 7.3 yields

$$
\begin{aligned}
x_c &= (C_1^2+C_2^2-C_3^2-C_4^2)\, g_x^*(\theta_1,\theta_2,\ldots,\theta_n;\ \mathbf{X},\mathbf{Y},\mathbf{Z}) + 2(C_2C_3+C_1C_4)\, g_y^*(\theta_1,\theta_2,\ldots,\\
&\qquad \theta_n;\ \mathbf{X},\mathbf{Y},\mathbf{Z}) + 2(C_2C_4-C_1C_3)\, g_z^*(\theta_1,\theta_2,\ldots,\theta_n;\ \mathbf{X},\mathbf{Y},\mathbf{Z}) + C_5 \qquad (7.10)\\
y_c &= 2(C_2C_3-C_1C_4)\, g_x^*(\theta_1,\theta_2,\ldots,\theta_n;\ \mathbf{X},\mathbf{Y},\mathbf{Z}) + (C_1^2-C_2^2+C_3^2-C_4^2)\, g_y^*(\theta_1,\theta_2,\ldots,\\
&\qquad \theta_n;\ \mathbf{X},\mathbf{Y},\mathbf{Z}) + 2(C_3C_4+C_1C_2)\, g_z^*(\theta_1,\theta_2,\ldots,\theta_n;\ \mathbf{X},\mathbf{Y},\mathbf{Z}) + C_6
\end{aligned}
$$

where

$$-C_5 = (C_1^2+C_2^2-C_3^2-C_4^2)\,x_0 + 2(C_2C_3+C_1C_4)\,y_0 + 2(C_2C_4-C_1C_3)\,z_0$$

(7.11) $$-C_6 = 2(C_2C_3-C_1C_4)\,x_0 + (C_1^2-C_2^2+C_3^2-C_4^2)\,y_0 + 2(C_3C_4+C_1C_2)\,z_0$$

If the goal is to estimate the camera-space kinematic relationships locally, that is to say, near pose A, of the cue itself, the estimation model of Eqs. 7.10 can be used directly together with samples. The latter could be tabulated, simultaneous samples of robot joint rotations together with camera-space location of the detected cue of interest.

As an example, consider the n = 3-degree-of-freedom robot pictured in *Figure 7-4b*. The nominal kinematics of the cue center are given by

(7.12)
$$x_P = [L_1\cos\theta_2 + L_2\cos(\theta_2 - \theta_3)]\cos\theta_1 - x_{O/A}$$
$$y_P = [L_1\cos\theta_2 + L_2\cos(\theta_2 - \theta_3)]\sin\theta_1 - y_{O/A}$$
$$z_P = L_1\sin\theta_2 + L_2\sin(\theta_2 - \theta_3) - z_{O/A}$$

where L_1 and L_2 are known link lengths and where $x_{O/A}$, $y_{O/A}$, $z_{O/A}$ are the known components of point o with respect to A. The exact value of these offsets used just to describe the nominal kinematics is unimportant; however it is advantageous and a very simple matter of altering judiciously $x_{O/A}$, $y_{O/A}$, $z_{O/A}$ to shift this frame of reference such that o is in the general vicinity of the positioned point at the desired maneuver terminus.

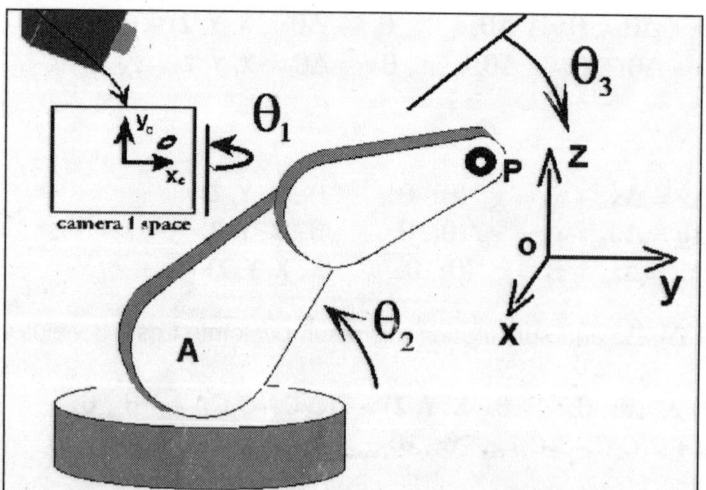

Figure 7-4b. The three-degree of freedom robot moves through a trajectory. At several stages of motion there is a simultaneous sampling of two things: The three joint angles and the camera's detection of its x_c and y_c coordinates of point P.

The present question is: What is the relationship between the current angular pose θ_1, θ_2, θ_3, and the position in camera space (x_c, y_c) of the cue center? In particular, our interest lies in this relationship *near* the previously mentioned "pose A", which we will take to be: θ_{1A} = 2.1 radians, θ_{2A} = 1.0 radians, θ_{3A} = 0.48 radians.

Consider the total of nine poses, or joint-angle combinations, as given in **Table 7-1**. Using a particular, assumed camera orientation and optical/electronic camera characteristics, we have also generated for this table nine simulated camera-space samples. The objective is to apply these samples to the estimation of camera-space kinematic relationships – algebraic relationships – capable of accurate prediction, in the vicinity of pose A, of cue locations as a function of future θ_1, θ_2, θ_3.

Table 7-1. Robot joint angles together with cue samples in 2D camera-space for use in calculation of camera-space kinematics. Item 9 is pose A.

ROBOT JOINT POSE (radians)			CAMERA SAMPLES (pixels)	
θ_1,	θ_2,	θ_3	x_c	y_c
1,0000	0.5000	0.1000	-205.47	96.38
1.2000	0.6000	0.2000	-137.91	103.80
1.4000	0.7000	0.3000	-63.55	98.45
1.6000	0.8000	0.4000	10.52	80.39
1.7000	0.9000	0.4000	43.21	35.42
1.8000	0.9500	0.4200	73.14	13.83
1.9000	0.9707	0.4400	102.12	2.98
2.0000	0.9866	0.4600	129.27	-7.62
2.1000	1.0000	0.4800	154.25	-18.84

The mathematical model for our camera-space kinematics estimates is Eqs. 7.10 with the functions g_x^*, g_y^* and g_z^* taking the form of x_P, y_P and z_P, respectively, of Eqs. 7.12. At this point in the discussion there is no need to specify **X**, **Y**, and **Z** of g_x^*, g_y^* and g_z^* since this initial example entails just a single cue, point P, with nominal forward kinematics given directly by Eqs. 7.12. The estimation model becomes:

$$f_x = (C_1^2+C_2^2-C_3^2-C_4^2)\ \{[\cos\theta_2+\cos(\theta_2-\theta_3)]\cos\theta_1\} + 2(C_2C_3+C_1C_4)$$
$$\{[\cos\theta_2+\cos(\theta_2-\theta_3)]\sin\theta_1 - 1\} + 2(C_2C_4-C_1C_3)\ \{\sin\theta_2+\sin(\theta_2-\theta_3) - 1\} + C_5$$

(7.13)

$$f_y = 2(C_2C_3-C_1C_4\ \{[\cos\theta_2+\cos\theta_2-\theta_3]\cos\theta_1 + (C_1^2-C_2^2+C_3^2-C_4^2)$$
$$\{[\cos\theta_2+\cos(\theta_2-\theta_3)]\sin\theta_1 - 1\} + 2(C_3C_4+C_1C_2)\ \{\sin\theta_2+\sin(\theta_2-\theta_3) - 1\} + C_6$$

where we have taken the nominal forward-kinematics length parameters to be $L_1 = L_2 = 1\text{m}$. Also we have taken the offsets to be: $y_{0/A} = z_{0/A} = 1\text{m}$; $x_{0/A} = 0$.

We think of Eqs. 7.13 as specifying the form of the camera-space location of the cue, x_c, y_c, as a function of the robot's three angles, θ_1, θ_2, θ_3, parameterized by the six "view parameters", $C_1 - C_6$. This may be written:

$$x_c = f_x(\underline{\boldsymbol{\theta}};\underline{\mathbf{C}}) \qquad (7.14)$$
$$y_c = f_y(\underline{\boldsymbol{\theta}};\underline{\mathbf{C}})$$

where $\underline{\boldsymbol{\theta}}$ includes implicitly all (three for this robot) elements of joint rotation and $\underline{\mathbf{C}}$ includes all elements of $C_1 - C_6$.

If we knew the elements of $\underline{\mathbf{C}}$, Eqs. 7.13 would be the algebraic camera-space kinematics expressions needed for CSM. Similar expressions for a second camera, adequately separated from this first camera, would in turn, as discussed in Chapter 6, allow for the solution of the joint pose $\underline{\boldsymbol{\theta}}$ needed to collocate point P with some physical juncture whose location in both cameras has been sampled and ascertained. We concentrate at the moment on applying the data presented in **Table 7-1** to produce best estimates of this "vector" $\underline{\mathbf{C}}$ in the first camera.

One way to achieve this objective is to select $\underline{\mathbf{C}}$ so as to minimize $J(\underline{\mathbf{C}})$ defined according to

$$J(\underline{\mathbf{C}}) = \Sigma_i \ [x_c^{\,i} - f_x(\underline{\boldsymbol{\theta}}^i;\underline{\mathbf{C}})]^2 + [y_c^{\,i} - f_y(\underline{\boldsymbol{\theta}}^i;\underline{\mathbf{C}})]^2 \qquad (7.15)$$

where Σ_i represents a summation over i in this case of nine terms representing the nine camera-space samples of **Table 7-1**, leading up to and including pose A. The symbols $x_c^{\,i}$, $y_c^{\,i}$, i=1,2, ... 9, are the elements of the ith camera-space sample itself, i.e. the two right columns of Table 7-1; and $\underline{\boldsymbol{\theta}}^i$, i=1,2, ... 9, are the three joint rotations of the ith sample.

There are several numerical methods for solving for the six elements of $\underline{\mathbf{C}}$ that minimize J of Eq. 7.15. Appendix A gives one of these. It is worth noting that due to the "nonlinearity" of the appearances of $\underline{\mathbf{C}}$ in our functions f_x and f_y, any of the methods will be "iterative" and also will present the possibility of convergence onto the wrong roots. These issues are easily handled in practice and discussed in Appendix A.

The key point here, however, can be seen in the plot of *Figure 7-5*. The model's best fits are not a perfect match for the data. Because this is a simulation that means

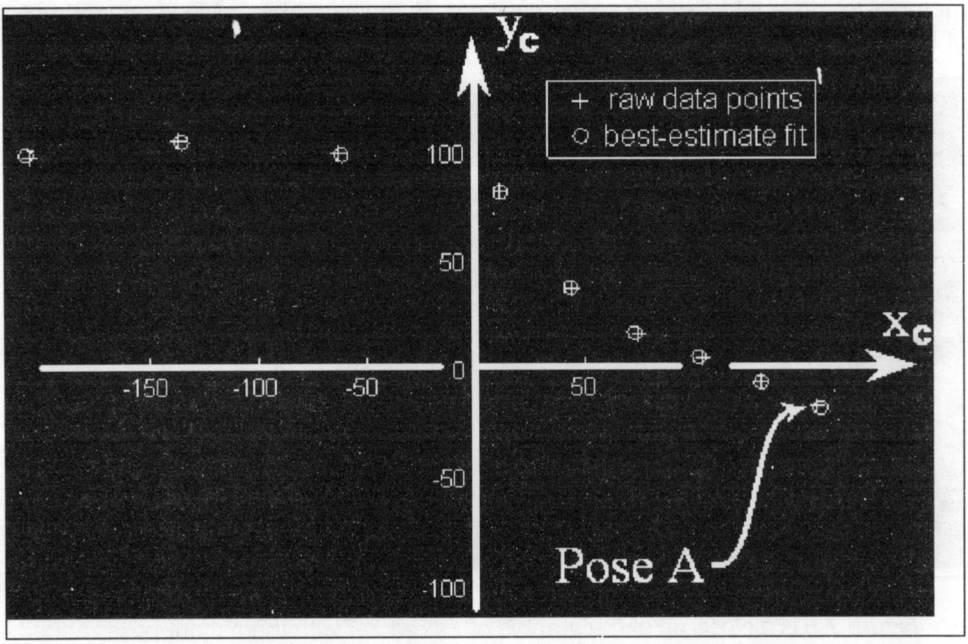

Figure 7-5. The "+" indicates camera-space samples of the cue, point P, of the robot of *Figure 7-4*. After minimization of $J(\underline{\mathbf{C}})$ as defined in Eq. 7.15, these best-fitting $\underline{\mathbf{C}}$ values were applied to the nine sets of joint rotations of Table 7-1 to determine the Eq-7.13 model's best fit of the data. The "O" icons represent these best fits.

that the model used to generate the data is not accommodated by the model used to fit the data. This reason for disparity, it turns out, is also the single biggest reason for a poor fit in most real-robot/optics instances. It is analogous to mapping large regions of the world using the flat-map assumptions (the distance between any two points on a map is proportional to their physical, over-land separation.) The model of Eqs. 7.13 presumes an "orthographic" mapping between three-dimensional physical space and two-dimensional camera space. Early paintings and works of art applied this framework; objects in the foreground were not painted larger, in accordance with perspective, than objects behind.

Just as the flat-earth mapping assumption would work well if all points on the map were in fact close together relative to the radius of the earth so too the orthographic mapping would work if the points fit to our model were closely spaced relative to their distance to the camera. The clearly imperfect "best fit" of *Figure 7-5* shows that that was not the case for this simulation. The more important indicator, however, has to do with predictions. Remember that the purpose of establishing locally the camera space kinematics is to be able to solve for *future* joint rotations that will cause the end member to move into camera-space positions as indicated by previous appearances of the workpiece or target object of manipulation.

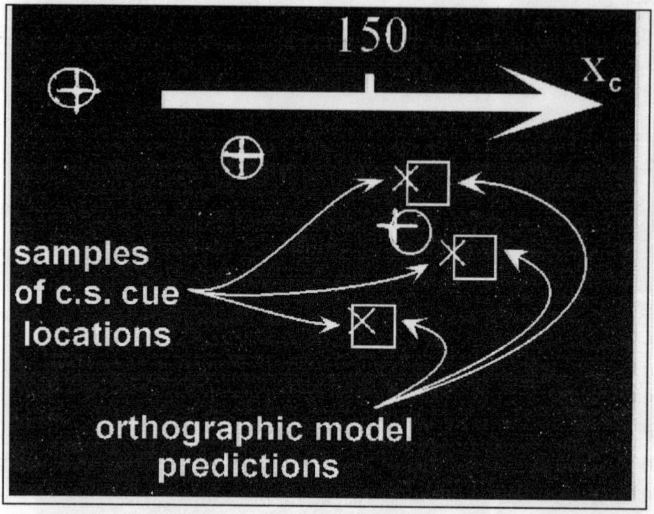

Figure 7-6. The three "**X**" icons represent samples of point P arrived at by adding 0.02 radians to one of each of the joint angles associated with pose A. This increment is added to a different angle for each of the indicated **X** marks. The corresponding box icons represent the model's predictions for each of these cases.

Figure 7-6 is a close-up of the graphic of *Figure 7-7* with three more poses added. Clustered around pose A are three more data samples represented by an "**X**" in *Figure 7-6.* These three were created from $\underline{\boldsymbol{\theta}}^9$, the joint coordinates of pose A, by adding 0.02 radians - in each case to one of the three elements of $\underline{\boldsymbol{\theta}}^9$. To assess the predictive ability of the model, the box icon illustrates the prediction of the model of Eqs. 7.13. Despite the close proximity of these points to pose A, which itself is one of the poses factored in to the $\underline{\mathbf{C}}$ estimates, the error in prediction is large – about one part in 50 of the full extent of camera space.

What can be done to improve this? An obvious answer, if it were a matter of applying the flat-earth model to a terrain map, would be to reduce the extent of data points factored into the mapping. The counterpart here would be to eliminate several of the data points furthest from pose A. Rather than do that directly, however, we do something nearly equivalent. We "deweight" those more distant points, giving them less influence in the determination of $\underline{\mathbf{C}}$.

Consider the following embellishment of the index of Eq. 7.15:

(7.16) $$J(\underline{\mathbf{C}}) = \Sigma_i \, \mathbf{W}_i \{[x_c^{\,i} - f_x(\underline{\boldsymbol{\theta}}^i;\underline{\mathbf{C}})]^2 + [y_c^{\,i} - f_y(\underline{\boldsymbol{\theta}}^i;\underline{\mathbf{C}})]^2\}$$

Finding the elements of $\underline{\mathbf{C}}$ that minimize this "weighted" alternative, not surprisingly, entails a procedure very similar to the case for Eqs. 7.15. It too is discussed in *Appendix A*.

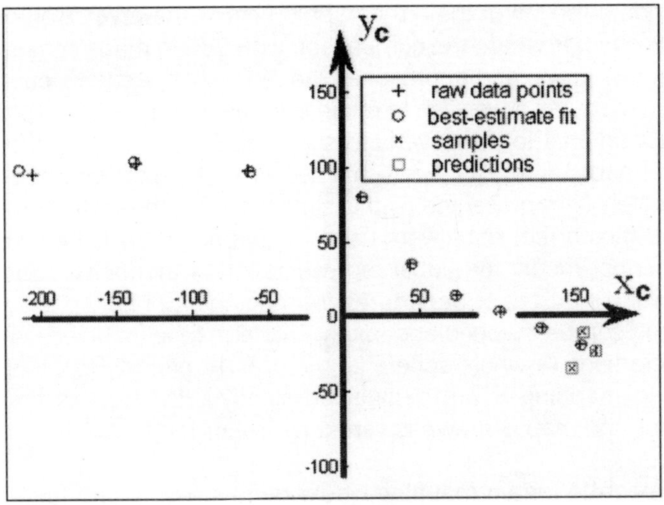

Figure 7-7. The same nine samples are again used, this time with the skewed-weighting best fits. The plot also shows the samples of *Figure 7-6* together with predictions based on skewed weight calculations for $\underline{\mathbf{C}}$.

Figure 7-7 shows a plot that differs from *Figure 7-5* in two respects: The minimized index $\mathbf{J}$ is that of Eq. 7.16 above, with $\mathbf{W}_i=i^2$, i=1,2, ... 9; and the three predictions of *Figure 7-8* are included, again clustered about pose A. This time it is clear that the three predictions are much better - an order of magnitude better, in fact - than the equal-weighting case. The effect of the highly skewed weighting ($\mathbf{W}_1$ is 1 whereas $\mathbf{W}_9$ is 81) is also clear on the superposition of best fits of the original nine points over their respective samples. Sample 1 to the extreme left of *Figure 7-7* is shown with a very large disparity relative to its best fit, and this disparity closes quickly with sample proximity to the most highly weighted pose A. A good exercise is to duplicate the plots of *Figures 7-5* and *7-8* using the data of Table 7-1. The vectors $\underline{\mathbf{C}}$ you should find are, for the unweighted (or equally weighted) case: $\underline{\mathbf{C}}$ = [0.4831 -0.1882 13.8063 -5.2318 7.8203 -5.7584]^T; and for the case of skewed weighting: $\underline{\mathbf{C}}$ = [-0.4693 0.1847 13.7289 -5.4968 -6.4356 -2.1593]^T.

Flattening

What is the benefit of skewing the weighting rather than merely truncating samples distant from pose A? One aspect of the answer to this has to do with the coarseness of vision and vision samples. Unlike our simulation above where each camera-space sample is virtually infinitely resolved, with no random error, *real* camera space is coarsely quantized. It is possible, and even the norm with laser-spot cues or the circular cues, to achieve sub-pixel accuracy in camera-space assessment of cues' center-point locations within an image, but even so each sample has comparatively little resolution or precision. Consequent numerical problems of convergence of our algorithm onto the correct $\underline{\mathbf{C}}$ vector are offset considerably by extending the tail of the data through deweighting – rather than truncation.

Better than simply extending the tail of the orthographic best fit, however, would be model improvement. This again is where the comparison with terrain maps comes in. As the extent of land mapped grows and approaches the radius of the earth's curvature, a flat map becomes inadequate if we wish to retain the rule: Distance between two points on the map is proportional to their overland separation. But drawing the map onto a spherical piece of paper allows for significant increase of the extent of the global surface that can accurately be represented by this simple rule. If the earth were a perfect sphere, in fact, and the correct scale were used for mapping, this spherical representation could be a perfect model for global representation of all points. But even with the reality of the world's bulges or departures from being a perfect sphere, comparatively large land areas can be mapped accurately onto our spherical piece of paper. Up to one or two steradians (a whole sphere is just over 12 steradians), this advantage or improvement in mapping is not diminished significantly even as the radius of the paper over which the map is drawn is varied by five or ten percent.

Another equivalent approach to terrain mapping would be to select a land juncture that represents the center of a flat map, but to apply a transformation such that departure of earth-surface junctures from that point carry with them departure from the simple mapping rule of proportionality. Such a map would look like a picture taken of a globe with the selected point closest to the camera. It would be an orthographic projection of the globe onto a flat surface. Land forms as you approach the outer rim of such a map would be deformed or distorted, and so the map may be of limited use to a human reader. (*Figure 7-8*) But a computer, with the mapping formula stored in memory, would have no problem transforming locations of two points on such a map to their overland separation across the face of the globe.

Figure 7-8. Projection of the globe onto a plane. The central juncture is the north pole, and there is little distortion of land masses near that juncture.

A similar accommodation to the camera mapping from physical space into camera space is what we here call "flattening." The kind of distortion that a viewer of an image treated by flattening would see amounts to elimination of the "perspective effect" where objects in the foreground appear larger compared to their actual size than do objects in the background. If the original image were a perfect perspective projection into a plane, and if flattening eliminated all perspective effects, then the model of Eqs. 7.13 would, with the right elements of $\underline{\mathbf{C}}$, perfectly express the mapping of the camera. Note that whereas terrain mapping of *Figure 7-8* is 2D to 2D, this flattening affects the way that we model and store the camera/lens mapping of *three-dimensional*, physical space into two-dimensional camera space. It is 3D to 2D.

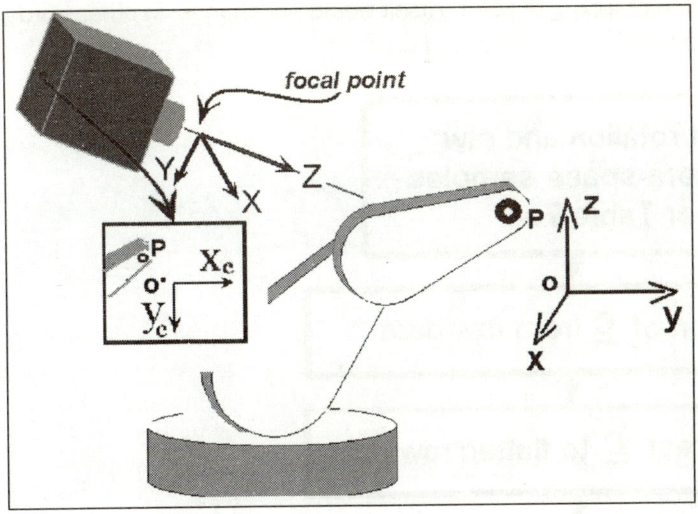

Figure 7-9. The perspective lens model is based upon the XYZ coordinate system with origin at the camera's focal point, and with the Z axis aligned with the camera's focal axis.

Consider the origin o relative to which the 3D robot-kinematics model, Eqs. 7.12, specify the forward kinematics of the robot of *Figure 7-9*. According to the simple perspective or pinhole camera model the coordinates x_c, y_c of point o will depend upon its location relative to the XYZ reference frame shown – the frame aligned with the camera – according to:

$$x_{co} = f\,X_o/Z_o$$

(7.17)

$$y_{co} = f\,Y_o/Z_o$$

where f is a constant and where X_o Y_o Z_o are the coordinates of o relative to the frame shown in *Figure 7-10* aligned with the camera. Similarly, the nine camera-space "samples" of Table 7-1 are simulated based upon this same model. Thus, for the ith sample, $x_{ci} = f\,X_i/Z_i$ and $y_{ci} = f\,X_i/Z_i$.

Camera-space coordinates of the ith sample, x^*_{ci} , y^*_{ci}, are said to be "flattened about point o" with application of the following conversion:

$$x^*_{ci} = x_{ci}\, Z_i/Z_0$$

(7.18)

$$y^*_{ci} = y_{ci}\, Z_i/Z_0$$

where x_{ci} , y_{ci} are the actual raw samples of the ith cue, each of which is indicated in *Figure 7-6* with a +. Note that the flattened samples represent an *orthographic* mapping of physical space since $x^*_{ci} = X_i$ x Constant, $y^*_{ci} = Y_i$ x Constant, where Constant = f/Z_0. Note further that the mapping of point o itself would not change; in other words $x_{co}^* = x_{co}$. Hence the phrase "flattening *about* point o".

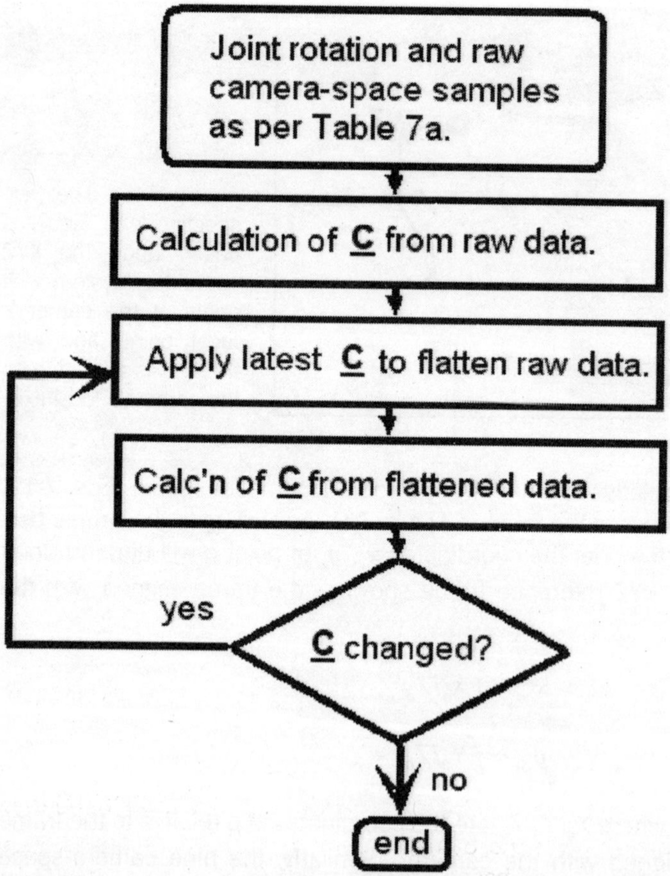

Figure 7-10. Flattening of raw camera-space data.

Because of this introductory use of simulated data, we know Z_0 exactly. In practice, however, this quantity would be only roughly known. Yet, as with drawing a terrain map onto a spherical surface, even with an error of ten percent or so, the benefit in terms of the extension or expansion of extent of the mapping is still substantially realized.

Application of Eqs. 7.18, then, begins with this assessed, approximate value of Z_0. The numerator Z_i, which is the Z component of position of the ith cue, like Z_0, is not known in advance. However, we do have the robot's forward kinematics model which nominally gives the position of the ith cue, that is the position of point P relative to point o when $\underline{\theta} = \underline{\theta}^i$. Letting ΔX_i ΔY_i ΔZ_i be the coordinates of point P relative to point o referred to the camera's XYZ coordinate frame, we know that there exists a 3x3 "direction cosine matrix" [c] that relates ΔX_i ΔY_i ΔZ_i to the $x_P(\underline{\theta}^i)$ $y_P(\underline{\theta}^i)$ $z_P(\underline{\theta}^i)$ of the forward kinematics model of Eqs. 7.12:

(7.19) $$[\Delta X_i \ \Delta Y_i \ \Delta Z_i]^T = [c]\,[x_P(\underline{\theta}^i)\ y_P(\underline{\theta}^i)\ z_P(\underline{\theta}^i)]^T$$

Furthermore, $Z_i = Z_0 + \Delta Z_i$. Thus, if we can identify the bottom row of [c] we can find a number for the value of Z_i needed in Eqs. 7.18 for flattening, i.e. $\Delta Z_i = c_{31}\, x_P(\underline{\theta}^i) + c_{32}\, y_P(\underline{\theta}^i) + c_{33}\, z_P(\underline{\theta}^i)$. An approximation to [c] can be found using the aforementioned vector $\underline{\mathbf{C}}$=[0.4831 -0.1882 13.8063 -5.2318 7.8203 -5.7584]^T, or the view parameters determined from the application of the raw camera-space samples to the orthographic model of Eqs. 7.13. At this point we consider the parameters $\underline{\mathbf{C}}$ based on equal weighting of all samples – those that gave rise to the "best fits" of *Figure 7-5.*

The first four elements of $\underline{\mathbf{C}}$ – C_1, C_2, C_3 and C_4 – relate to the four elements of "Euler parameters", e_0, e_1, e_2, e_3, a particular form of quaternion, according to:

$$e_{i-1}=C_i/[C_1^2+C_2^2+C_3^2+C_4^2]^{1/2}, \quad i=1, 2, 3, 4.$$

These in turn relate to our elements of interest of [c] according to:

(7.20) $$\begin{aligned} c_{31} &= 2(e_1e_3 + e_0e_2) \\ c_{32} &= 2(e_2e_3 - e_0e_1) \\ c_{33} &= e_0^2 - e_1^2 - e_2^2 + e_3^2 \end{aligned}$$

The reader can verify that, for the equal-weight case of *Figure 7-5*, the direction cosine elements of Eqs. 7.20 result in: c_{31} = 0.0701389, c_{32} = -0.661078, c_{33} = -0.747032. These numbers in turn produce flattened data x^*_c, y^*_c as indicated in

Table 7-2. Based on the sequence shown above to produce the first round of **C** using equal weighting of data from **Table 7-1**, a new **C** can be produced using equal weighting with the once-flattened data of **Table 7-2**.

....

Table 7-2. After the initial round of "flattening" about point *o* of *Figure 7-9* we have the indicated camera-space values.

ROBOT JOINT POSE (radians)			CAMERA SAMPLES (pixels after one round of flattening)	
θ_1,	θ_2,	θ_3	x^*_c	y^*_c
1,0000	0.5000	0.1000	-198.35	93.03
1.2000	0.6000	0.2000	-128.63	96.82
1.4000	0.7000	0.3000	-57.93	89.71
1.6000	0.8000	0.4000	9.49	72.50
1.7000	0.9000	0.4000	38.61	31.65
1.8000	0.9500	0.4200	65.38	12.36
1.9000	0.9707	0.4400	91.67	2.68
2.0000	0.9866	0.4600	116.78	-6.88
2.1000	1.0000	0.4800	140.52	-17.17

....

Once these new view parameters **C** are found, a chart similar to that of *Figure 7-5* could be constructed superimposing the best fit of the nine entries of **Table 7-2** onto the once-flattened data themselves. The result would be a great improvement of the fit over that of *Figure 7-5*, but still not a perfect fit. This being a simulation, where the only difference between the estimation model and the model used to create the data is the effect of perspective, there is the natural question as to why it is that flattened data – data deliberately altered to be compatible with an orthographic mapping - are not in fact perfectly accommodated by the orthographic model. The reason lies in the imperfection of the first round of flattening itself: Values of c_{31} c_{32} c_{33} used to convert physical-space coordinates to the reference frame of the camera were based upon an initial **C** that represented the best fit of an imperfect model. This orthographic model will "absorb" those imperfections resulting in geometric inferences that are a bit off.

Nevertheless, the once-flattened data are closer to conformity with the orthographic ideal. As a consequence, the new set of direction-cosine elements c_{31} c_{32} c_{33} resulting from this iteration's **C** will be closer to the truth. *Figure 7-10* illustrates the iterative algorithm that, for the present case of simulated data, results in the perfect fits of *Figure 7-11*. The exact value of Z_0 for this case is known to be 3.0 meters.

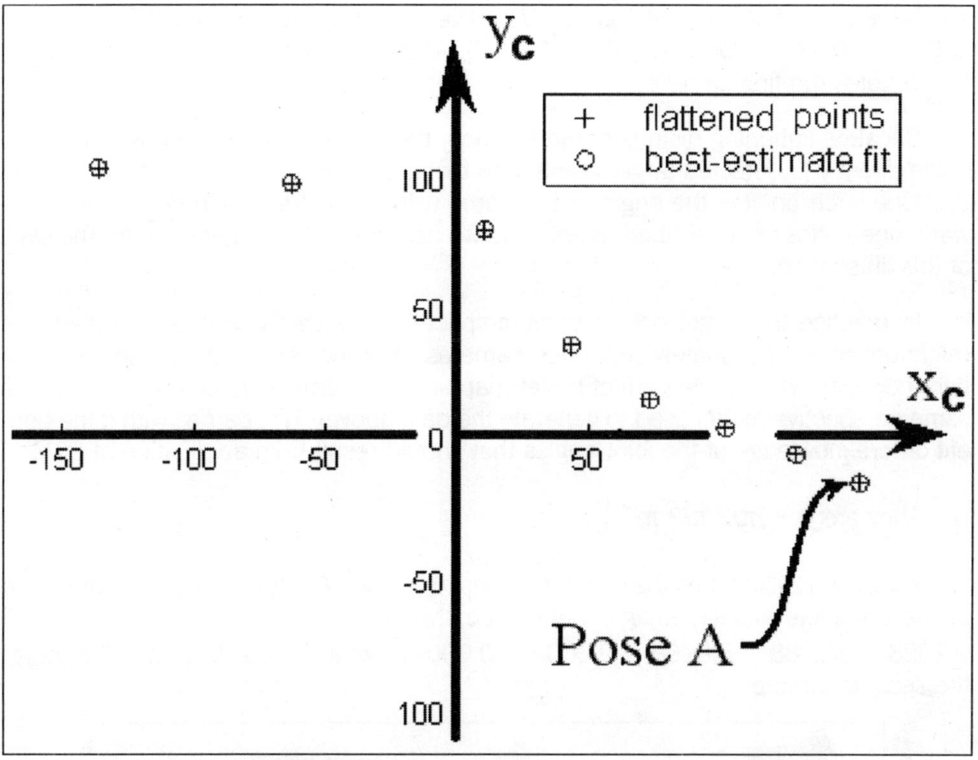

Figure 7-11. After the iterative procedure of *Figure 7-10* is complete, the resulting camera-space data are perfectly fit by the orthographic camera model. Perfection of the fit is a consequence of the ideal pinhole model used to simulate the original data and would not be affected by a skewed weighting scheme.

With real data it is common that flattening will eliminate most of the error of any fit. This is because, while the pinhole model does not characterize perfectly real lenses, it does capture the biggest disparity between the orthographic assumption of the estimation model and actual lens mapping. This is true even if the assessed Z_0 differs from the true value on the order of ten percent. It is similar to the improvement in use of the simple terrain mapping assumption that results from plotting maps onto a spherical surface. Even though the radius of the sphere may be off by ten percent compared to the scaling of the map, and even though the perfect sphere is not an exact representation of the imperfectly spherical earth, the improvement can be very significant. Even with the use of flattening, skewed weighting in favor of samples closer to the terminus prior to prediction should still be used, as with *Figure 7-6*, but importantly: flattening results in the ability to make the skewness of this weighting much more gradual. That in turn removes the numerical dangers of working with noisy camera-space data. The reader is encouraged to complete the iterative process and so replicate the result of *Figure 7-11*. The converged vector $\underline{\mathbf{C}}$ used in this figure is $\underline{\mathbf{C}}_1$ = [-0.0000 0.0000 13.0656 -5.4120 -0.0004 0.0001]^T, where the subscript "1" denotes the first camera.

Because flattening distorts camera space, the above vector $\underline{\mathbf{C}}$ used is only applicable directly to detected target points with distance from the camera's focal point of Z_0. One such point is the origin of the frame with respect to which our nominal forward kinematics are described, point o; so we use this as the target point for the sake of this illustration.

In practice the target point's camera-space coordinates would be sampled in a minimum of two adequately separated cameras. For the sake of the present illustration, however, we assume perfect target mapping into camera space according to the same perspective model used to generate the data above. This carries with it the benefit of foreknowledge of the joint angles that should result from application of CSM.

They are $\underline{\theta} = [\pi/2\ \pi/2\ \pi/2\ ]^T$.

Figure 7-12 illustrates the second camera in place. Applying the procedure given above, complete with flattening about point o, the elements of $\underline{\mathbf{C}}$ become: $\underline{\mathbf{C}}_2$ = [3.8268 9.2388 9.2388 3.8268 -0.0000 0.0000]^T, where the subscript "2" denotes the second camera.

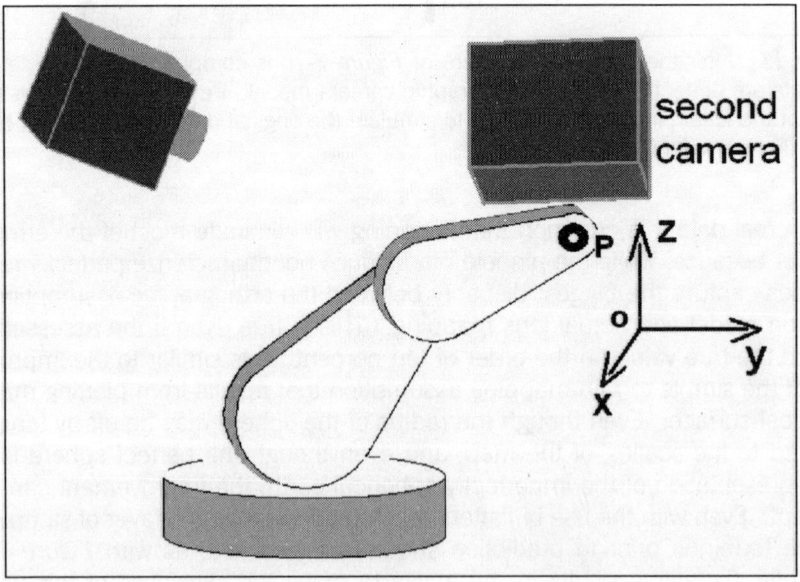

Figure 7-12. A second camera is added in order to solve for the joint rotations needed to collocate point P with point o in each of two camera spaces - and hence also in physical space.

Because this is a simulation we can determine perfectly the target camera-space locations that would correspond with the origin o of the coordinate frame x-y-z of *Figure 7-12*. In Camera 1, that juncture would be $(x_{c1}{}^t\ y_{c1}{}^t)$=(0.0 0.0), where the superscript t is for "target". In Camera 2, it is for this simulation also $(x_{c2}{}^t\ y_{c2}{}^t)$=(0.0 0.0). For a physical positioning problem these target camera-space positions would be based upon camera samples of the target body, for example using laser spots and image differencing as discussed in Chapter 6. It is important to note that, because we have flattened about the target point o it is permissible to use for the camera-space kinematics relationships $f_{1x}(\theta_1, \theta_2, \theta_3)$ and $f_{1y}(\theta_1, \theta_2, \theta_3)$, $f_{2x}(\theta_1, \theta_2, \theta_3)$ and $f_{2y}(\theta_1, \theta_2, \theta_3)$ of Chapter 6, the view parameters $\underline{\mathbf{C}}_1$ and $\underline{\mathbf{C}}_2$ given immediately above. Referring to Eqs 7.14, we therefore have for the position of point P:

(7.21)

$$\begin{aligned} f_{1x}(\theta_1, \theta_2, \theta_3) &= f_x(\underline{\boldsymbol{\theta}};\underline{\mathbf{C}}_1) \\ f_{1y}(\theta_1, \theta_2, \theta_3) &= f_y(\underline{\boldsymbol{\theta}};\underline{\mathbf{C}}_1) \\ f_{2x}(\theta_1, \theta_2, \theta_3) &= f_x(\underline{\boldsymbol{\theta}};\underline{\mathbf{C}}_2) \\ f_{2y}(\theta_1, \theta_2, \theta_3) &= f_y(\underline{\boldsymbol{\theta}};\underline{\mathbf{C}}_2) \end{aligned}$$

Applying again the techniques outlined in Appendix A, we can solve for the values of θ_1, θ_2, θ_3 that minimize

(7.22)

$$J(\theta_1, \theta_2, \theta_3) = [x_{c1}{}^t - f_{1x}(\theta_1, \theta_2, \theta_3)]^2 + [y_{c1}{}^t - f_{1y}(\theta_1, \theta_2, \theta_3)]^2 + [x_{c2}{}^t - f_{2x}(\theta_1, \theta_2, \theta_3)]^2 + [y_{c2}{}^t - f_{2y}(\theta_1, \theta_2, \theta_3)]^2$$

It is a good exercise to verify the expectation that minimization of J of Eq. 7.22 results in $\theta_1 = \pi/2$, $\theta_2 = \pi/2$, $\theta_3 = \pi/2$. Interestingly, there are three other possible results: $\theta_1 = -\pi/2$ $\theta_2 = 0$ $\theta_3 = \pi/2$ and $\theta_1 = -\pi/2$ $\theta_2 = \pi$ $\theta_3 = \pi/2$ and $\theta_1 = \pi/2$ $\theta_2 = -\pi/2$ $\theta_3 = -\pi/2$. These are all minima, and they all result in the minimum possible value J=0. This is common in robotics. It is not in practice a problem because the roots to which the algorithm of Appendix A converges depend upon the initial guess. This will ordinarily be a guess that is near the pose from which the terminus is approached.

It is also possible to converge onto a nonglobal, or local minimum. Or a local or global maximum; or a "saddle point". All of these cases can be recognized and eliminated in software, in the writers' experience, for the present formulation.

In a real maneuver, as opposed to this simulation, flattening about the target point requires a nesting of iteratively improving estimates of the target point about which flattening occurs - within a loop. In the context of the example above, this would entail solution for θ_1, θ_2, θ_3 based upon unflattened **C** estimates in the two cameras' detected target points x_{c1}^t y_{c1}^t x_{c2}^t y_{c2}^t, and minimization of J of Eq. 7.22 to find θ_1, θ_2, θ_3. With an initial θ_1, θ_2, θ_3 in hand and an assumed Z_o, or focal-axis component between camera and origin, the data-flattening step of Eqs. 7.18, can be executed initially

(7.23)

$$x_{ci}^* = x_{ci}\,(Z_0 + \Delta Z_i)/(Z_0 + \Delta Z_t)$$

$$y_{ci}^* = y_{ci}\,(Z_0 + \Delta Z_i)/(Z_0 + \Delta Z_t)$$

where ΔZ_t is the component of displacement along the focal axis of the camera in question of the initial estimate of the target point relative to the origin, point o of *Figure 7-12.* This is found from the initial values of θ_1, θ_2, θ_3 using the nominal forward kinematics model of Eqs. 7.12. Denoting by x_t, y_t, z_t, the nominal target-point location in the 3D reference frame of the robot, we have:

(7.24)

$$x_t = [L_1\cos\theta_2 + L_2\cos(\theta_2 - \theta_3)]\cos\theta_1 - x_{o/A}$$

$$y_t = [L_1\cos\theta_2 + L_2\cos(\theta_2 - \theta_3)]\sin\theta_1 - y_{o/A}$$

$$z_t = L_1\sin\theta_2 + L_2\sin(\theta_2 - \theta_3) - z_{o/A}$$

In accordance with Eq. 7.18 above:

(7.25)

$$[\Delta X_t\ \Delta Y_t\ \Delta Z_t]^T = [c]\ [x_t\ \ y_t\ z_t]^T$$

The needed bottom row of [c] is found as per the procedure outlined above for flattening about point o, and $\Delta Z_t = c_{31}\, x_t + c_{32}\, y_t + c_{33}\, z_t$. As before: $\Delta Z_i = c_{31}\, x_P(\underline{\theta}^i) + c_{32}\, y_P(\underline{\theta}^i) + c_{33}\, z_P(\underline{\theta}^i)$. The first approximation to [c] is, for any target point, found using the unflattened $\underline{\mathbf{C}}_1$= [0.4831 -0.1882 13.8063 -5.2318 7.8203 -5.7584]^T, or the view parameters determined from the application of the raw camera-space samples to the orthographic model of Eqs. 7.13. These parameters $\underline{\mathbf{C}}_1$ are based on equal weighting of all samples – those that gave rise to the "best fits" of *Figure 7-5.*

With once-flattened data in hand, a new set of $\underline{\mathbf{C}}_1$ and $\underline{\mathbf{C}}_2$ are found. It would, at this point, be possible to continue iteration of **C** through to convergence, as per the first example. However, it is important to note that the target point about which flattening would then be conducted is in error, since it has been based upon a solution of θ_1, θ_2, θ_3 using $\underline{\mathbf{C}}_1$ and $\underline{\mathbf{C}}_2$ calculated from a faulty orthographic assumption. The once-iterated-upon $\underline{\mathbf{C}}_1$ and $\underline{\mathbf{C}}_2$, however, represent an improvement over the strict orthographic assumption. It is therefore preferable to use these to reestablish x_t, y_t, z_t prior to the next upgrade of $\underline{\mathbf{C}}_1$ and $\underline{\mathbf{C}}_2$.

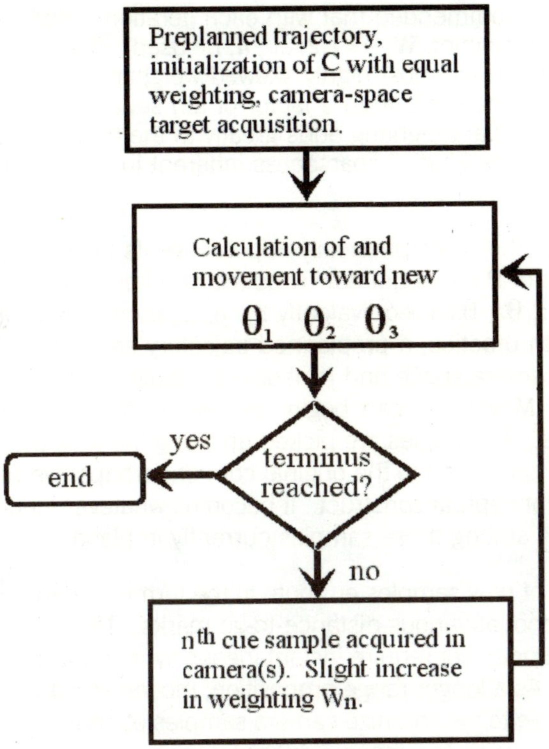

Figure 7-13. Flow chart of movement toward target while updating terminal joint coordinates en route. Data from the preplanned or initialization trajectory can be saved and applied to several maneuvers.

This procedure, outlined in the flowchart of *Figure 7-13*, converges reliably and yields the best current estimate of our terminal joint rotations θ_1, θ_2, θ_3 or their equivalent via Eqs. 7.12, x_t, y_t, z_t. It is important to note that equivalency between θ_1, θ_2, θ_3 and x_t, y_t, z_t is due to the nominal kinematics model of Eqs. 7.12. We are not saying that model is physically correct, or that x_t, y_t, z_t would be a good actual measure of the position of the target point relative to the physical coordinate frame with respect to which the nominal kinematic model of the robot is specified. That is virtually always, in fact, not the case. The frame relative to which the estimates x_t, y_t, z_t *would* be accurate is that which is mentioned earlier - the frame of a kinematically perfect robot if, at θ_1, θ_2, θ_3 or "pose A," its end effector were physically present in the same location as the end effector of the actual, imperfect robot posed in the same joint configuration. This physical frame shifts as the real robot occupies different regions of its workspace. Fortunately, with manipulation in camera space we never have to keep explicit track of this frame's shifting about.

As indicated in *Figure 7-13*, it is recommended that with each iteration a slightly more skewed weighting scheme for determining $\mathbf{W}_i$, $i=1,2,\dots,n$, be used. The application of flattening means that even the final, most highly skewed weighting can still be comparatively gradual, with perhaps no more than a factor of ten separating $\mathbf{W}_1$ from $\mathbf{W}_n$. The ability to use such a gradual weighting adds greatly to the robustness and precision of this overall scheme to the kind of coarseness inherent in visual cue samples.

It is important to note too that, as the robot physically approaches its target, the most valuable camera-space samples of **P** are acquired. Thus the number n of samples factored into the estimates of θ_1, θ_2, θ_3 or equivalently x_t, y_t, z_t increases with advancement toward the objective. In practice, a preplanned trajectory encompassing a wide range of physical space, camera space and joint space should be used to establish a core group of samples. Movement can begin, as per the flowchart of *Figure 7-13*, using just these samples. As samples are picked up along the way, they are given more weight, progressively, compared to the original core of perhaps twelve samples. "Pose A" is thus a kind of conceptual construct. It becomes whatever juncture or pose is closest to the terminus among those samples currently in place.

As a practical matter, acquisition of new samples en route to the terminus can be achieved by physically pausing the robot at various distance-to-go marks. This is the technically easiest solution because it does not require instantaneous synchronization of image and joint-rotation sampling. As a longer-range proposition, motion should be continuous. It is not difficult in principle to synchronize camera samples of end member cues with joint-rotation samples even as the maneuver based upon pre-current-last-sample estimates of the terminus ensues. In fact today's fast computers would allow in general for a great many such samples to be factored in to the terminal joint rotations, even during relatively quick robot motion. Moreover, there are a number of ways to queue intermediate robot joint-level reference inputs in such a way as to accommodate smoothly small updates in the terminal joint-pose calculations. That these updates are delayed relative to the point in time where the most recent camera/joint samples that factored into their calculation occurs is not, for a holonomic robot, a problem.

Camera-space-kinematics estimation of a point not detectable by cameras

There are few tasks that can be achieved based on the above sequential sampling of the same end-member juncture that is also placed or collocated with a surface juncture. At the same time there are few tasks that *cannot* be achieved if accurate camera-space kinematics can be estimated using samples of end-member junctures – or cues – away from the positioned points. So estimation of camera-space kinematics for this case is especially important.

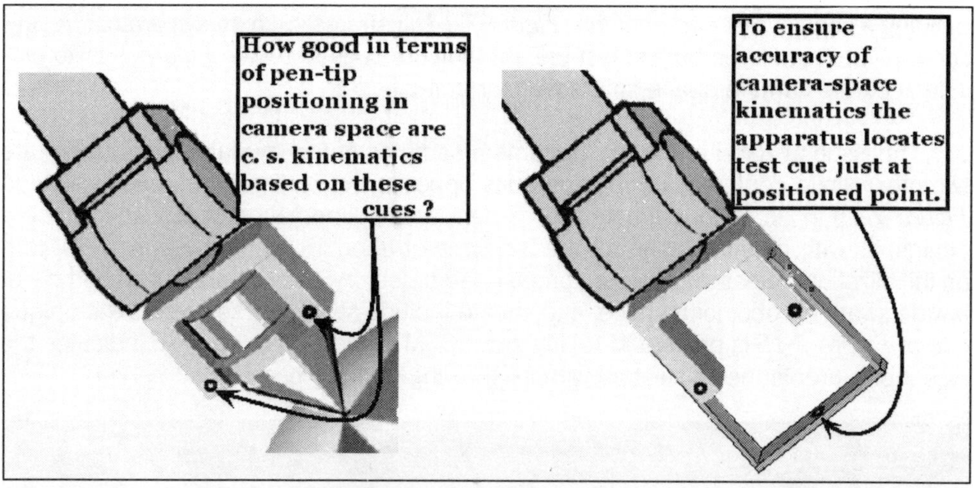

Figure 7-14. To verify directly the ability of the cues available on the tool of interest (to the left) to collocate with a selected surface point the pen tip, a special test apparatus can be built. Applying the same sampling/estimation regimen that is designed for the actual task, the apparatus to the right allows for verification of the camera-space location of a point located identically with respect to the sampled cues as the pen tip of the actual tool.

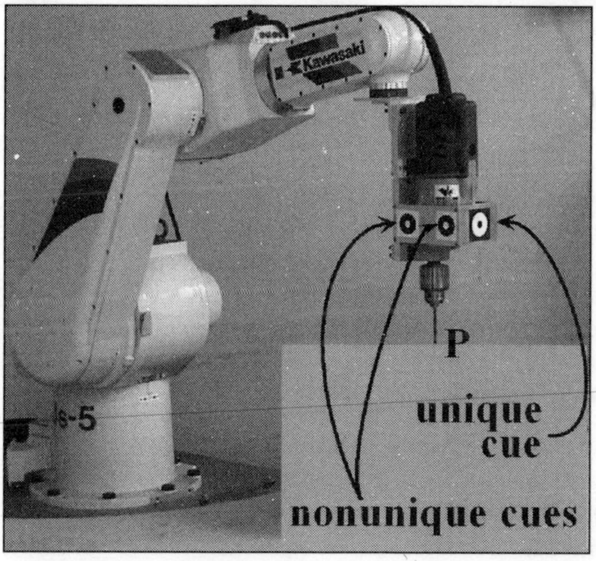

Figure 7-15. Cues detected in camera space are utilized to determine the camera-space kinematics of point **P**, the point that will be collocated with a selected surface juncture. As discussed in Chapter 8, the process of drill-bit entry can be defined in terms of a gradual shifting of this point **P** along the axis of the drill bit.

In devising a camera-space-manipulation system, it is important to ascertain with a high level of certainty that the sampling and estimation scheme results in accurate camera-space kinematics estimates for the positioned point(s) of interest. Referring for example to the pen-plotter of *Figure 7-1*, this accuracy can be verified directly by building a special *test end effector*. *Figure 7-14* illustrates such an apparatus. A similar apparatus could be built to test the ability of the cues shown in *Figure 7-15* to provide accurate camera-space kinematics for drill-bit tip **P**.

The issue at hand is the use of camera samples of five available cues (there are two other white-centered, nonunique cues opposite those nonunique cues visible in *Figure 7-15* in order to estimate the camera-space kinematics of **P**. This process begins, as with the simulation for the 3-degree-of-freedom robot of *Figure 7-4a*, with an initialization trajectory. Such a "preplanned trajectory" is arbitrary but should cover a wide swath of robot joint space, and should include several poses where the unique cue of *Figure 7-15* is presented to any given CSM camera. *Figure 7-16* indicates the extent of a preplanned trajectory with its showing of three poses used herein.

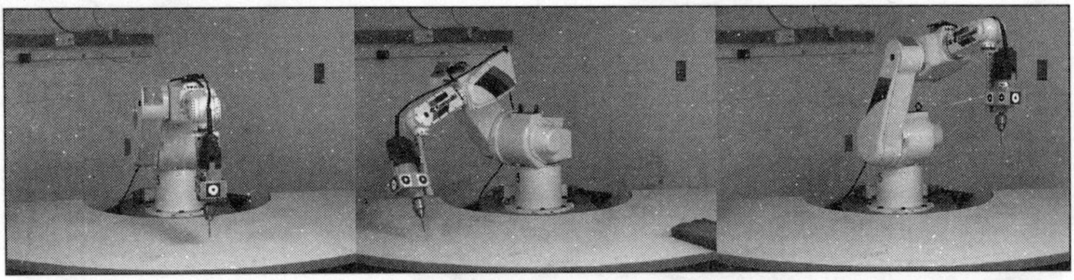

Figure 7-16. The preplanned trajectory should include several poses where a breadth of end-effector positions is presented to all cameras. It is necessary that the unique cue be detectable in many of the poses for each of the participant cameras. Other than that requirement, the choice is arbitrary, although consecutive poses should be contiguous.

Table 7-3 lists the camera-space samples for one of the participant cameras, together with corresponding joint rotations. Note that the number of cue detections is variable. Note further that only the unique cue's identity is certain. Yet, it is important to apply all cue data to the estimates of camera-space kinematics, f_x f_y, of point **P**, particularly since, as indicated in *Figure 7-15*, **P** is rather removed from the cues. This separation of the positioned point **P** from visually accessible cues makes all the more critical the improved modeling associated with flattening.

The flattening procedure begins very similarly with that of the single cue above. In particular, we apply the single, unique cue, and its appearances in each participant CSM camera to compute an equally weighted set of parameters $\underline{\mathbf{C}}_i$, i=1,2, ... n_{cam}, where n_{cam} is the number of cameras.

Because there are several points of interest on the end effector – including all the cues and each juncture P defined during evolution of the approach – the definition of the nominal kinematics functions is now amplified. Consider the form of nominal forward kinematics of Eqs. 7.6 above, repeated here as Eqs. 7.26:

Table 7-3. These are the data for the drilling task that have been compiled from a preplanned trajectory - for one particular camera. Note that the number and type of cue detected is variable camera by camera.

ROBOT JOINT ROTATIONS (degrees)						CAMERA SPACE CUES coordinates (pixels)			
						unique		non-unique	
θ_1	θ_2	θ_3	θ_4	θ_5	θ_6	x_c	y_c	x_c	y_c
-34.14	53.39	-58.19	10.04	-54.44	8.43	253.00	28.00	209.70	33.47
								186.42	20.36
-22.69	53.39	-58.18	-2.83	-54.44	8.44	134.82	64.60	124.65	63.55
								101.87	51.75
5.43	51.58	-61.87	-2.83	-48.71	-3.84	-160.21	79.50	-196.84	51.15
								-161.10	40.23
21.90	30.47	-77.42	10.03	-54.45	-55.17	-205.92	-54.71	-297.40	-66.61
								-254.52	-73.56
0.85	27.75	-80.69	-7.13	-54.45	-23.78	-74.33	-39.53	-34.92	-50.19
								-78.50	-56.80
-21.22	27.75	-84,38	-7.13	-54.45	5.12	85.96	-44.18	130.50	-60.63
								87.68	-63.31
-46.64	16.18	-77.01	-7.12	-60.18	53.95	208.40	-75.04	273.70	-91.13
								234.21	-95.45
-46.64	16.18	-77.01	17.21	-60.18	50.15	234.20	-189.08	285.07	-223.05
								243.70	-213.50
-23.94	16.18	-77.01	-7.13	-60.18	8.94	105.33	-166.50	147.31	-148.81
								103.20	-147.92
5.01	16.18	-82.33	-7.33	-60.18	-55.98	-66.39	-117.89	-96.73	-121.00
								131.41	-130.22
32.10	12.55	-82.33	-7.13	-60.18	-55.98	-285.36	-165.36		
32.10	28.43	-97.47	-1.41	-60.18	-55.98	-227.15	-37.47		
-28.31	34.78	-88.88	-1.40	-60.18	21.23	99.00	0.50	147.76	-19.36
								109.88	-31.00

$$x = g_x^*(\theta_1, \theta_2 \ldots, \theta_n; \mathbf{X}, \mathbf{Y}, \mathbf{Z})$$
$$y = g_y^*(\theta_1, \theta_2 \ldots, \theta_n; \mathbf{X}, \mathbf{Y}, \mathbf{Z}) \quad (7.26)$$
$$z = g_z^*(\theta_1, \theta_2 \ldots, \theta_n; \mathbf{X}, \mathbf{Y}, \mathbf{Z})$$

The coordinates **X**, **Y**, **Z** refer to the coordinate system fixed to the end effector. Every cue and every point P defined would therefore have a known and permanent **X**, **Y**, **Z**. A standard way in robotics of expressing the nominal forward kinematics relationships of Eqs. 7.26 is provided in Appendix B. Using the nominal forward kinematics of the robot of *Figure 7-15* together with the preplanned data of **Table 7-3** results in the chart shown in *Figure 7-17*. Note that, as with the chart of *Figure 7-5*, there is no flattening involved. Because the cameras are, for this case, close to the robot in comparison to the span or extent of the preplanned trajectory itself, the orthographic model results in a highly imperfect ability to fit the data. It is important to note that the view parameters, for a given set of data, will possibly converge to any one of perhaps three or four different values of **C** depending upon the initial guess used in the iterative procedure of Appendix A. Inspection of the plot of *Figure 7-17* reveals that, while the fit is imperfect, it is broadly consistent with the raw data; that is an indication that the *desired* convergence has been reached.

The first use that can be made of these view parameters is identification of the nonunique cues listed in **Table 7-3**. A preferred weighting of samples acquired at or near the pose which is under consideration is used. This allows for nonunique-cue identification despite the relatively poor overall fit. The critical dimension for determining this matter is the physical separation of the nonunique cues as pictured in *Figure*

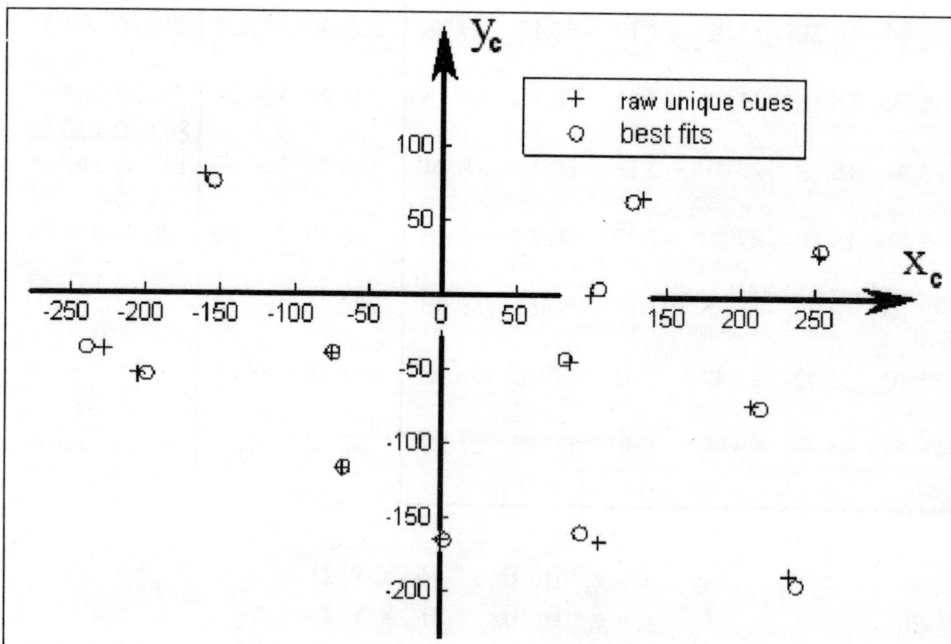

Figure 7-17. Raw data and best fit of preplanned-trajectory poses - unique cue only.

7-15. The closer they are to one another the more accurate the pose-by-pose predictions must be.

Also, it should be noted that in addition to the two nonunique cues visible in *Figure 7-15*, there are also two occluded cues of the same type. Rejection of them in identifying a given nonunique-cue indication in **Table 7-3** can be achieved using a logic based upon the equal-weight, unique-cue-based view-parameter estimates **C** mentioned above. In particular, recall that the first four elements of **C** relate to the Euler parameters that in turn describe the orientation of the camera's **Z** axis with respect to the robot's base coordinates, in accordance with

$$e_{i-1}=C_i/[C_1^2+C_2^2+C_3^2+C_4^2]^{1/2}, \quad i=1, 2, 3, 4. \text{ and Eqs. 7.20.}$$

$$c_{31} = 2(e_1e_3 + e_0e_2)$$
$$c_{32} = 2(e_2e_3 - e_0e_1)$$
$$c_{33} = e_0^2 - e_1^2 - e_2^2 + e_3^2$$

where c_{31} c_{32} c_{33} represent the three elements of the bottom row of the direction cosine matrix [c] which relates the robot frame to the camera-fixed physical coordinate system **X Y Z** as pictured in *Figure 7-9*. These elements can be understood in terms of a unit vector **n** that is normal to and directed away from the surface of a candidate nonunique cue, as pictured in *Figure 7-18*.

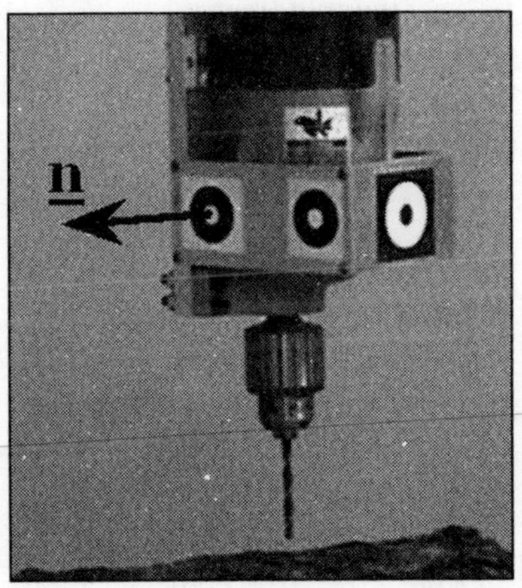

Figure 7-18. Is the cue with unit normal **n** facing toward or away from the camera of interest? If facing away, we can reject it apriori as a candidate cue identity for that camera/pose combination in question.

(7.27) $$[n_X\ n_Y\ n_Z]^T = [c]\ [n_x\ n_y\ n_z]^T$$

where n_X n_Y n_Z and n_x n_y n_z are the components of **n** with respect to the camera-fixed and robot-base-fixed frames, respectively. Our interest lies with n_Z for the particular cue/pose combination of interest. Of course the calculation of n_Z from the bottom-most equation of Eqs. 7.27, based as it is on several approximations, is itself only approximate. Nominally, we consider the cue to be "facing away from the camera" if $n_Z > 0$. Actually, we adopt the more conservative if somewhat arbitrary condition of rejecting the cue if $n_Z > -0.1$. This compensates for the many approximations involved. Also a case of $n_Z = 0$ means nominally that the camera's focal axis is parallel to the plane of the cue. As a practical matter, the cue-detection algorithm of *Appendix C* will not register an actual cue unless and until that angle departs more than 20 degrees or so from this case.

Computation of $n_Z = c_{31}n_x + c_{32}n_y + c_{33}n_z$ of course also requires the current-pose assessment of $n_x\, n_y\, n_z$. This can be achieved in a variety of ways. Perhaps the easiest way conceptually is to use the nominal forward-kinematics model of Appendix B defining two points, $\mathbf{P}_1$ and $\mathbf{P}_2$, separated by one unit of length, as shown in *Figure 7-19*. Denoting the end-member-frame coordinates of $\mathbf{P}_1$ and $\mathbf{P}_2$ according to $\mathbf{X}_1\ \mathbf{Y}_1\ \mathbf{Z}_1$ and $\mathbf{X}_2\ \mathbf{Y}_2\ \mathbf{Z}_2$, respectively, $n_x = g_x{}^*(\theta_1, \theta_2 \ldots, \theta_n; \mathbf{X}_2, \mathbf{Y}_2, \mathbf{Z}_2) - g_x{}^*(\theta_1, \theta_2 \ldots, \theta_n; \mathbf{X}_1, \mathbf{Y}_1, \mathbf{Z}_1)$, $n_y = g_y{}^*(\theta_1, \theta_2 \ldots, \theta_n; \mathbf{X}_2, \mathbf{Y}_2, \mathbf{Z}_2) - g_y{}^*(\theta_1, \theta_2 \ldots, \theta_n; \mathbf{X}_1, \mathbf{Y}_1, \mathbf{Z}_1)$, and $n_z = g_z{}^*(\theta_1, \theta_2 \ldots, \theta_n; \mathbf{X}_2, \mathbf{Y}_2, \mathbf{Z}_2) - g_y{}^*(\theta_1, \theta_2 \ldots, \theta_n; \mathbf{X}_1, \mathbf{Y}_1, \mathbf{Z}_1)$.

With the above approach applied to automatic rejection of cues facing away from the camera, we can begin identification of the remaining nonunique cues using a weighted version of the estimates $\underline{\mathbf{C}}$ based upon the unique cue only. Consider the fifth entry of the preplanned trajectory of **Table 7-3**. The detected camera-space coordinates of the unique cue in the fifth pose is presented together with the equal-weight, unflattened best fit in a magnified way in *Figure 7-20*. Added to these points are the actual detected locations of two white-centered, nonunique cues in the fifth pose. Finally, based upon the equally weighted $\underline{\mathbf{C}}$ given above, the predictions of these two cues' camera-space locations are shown.

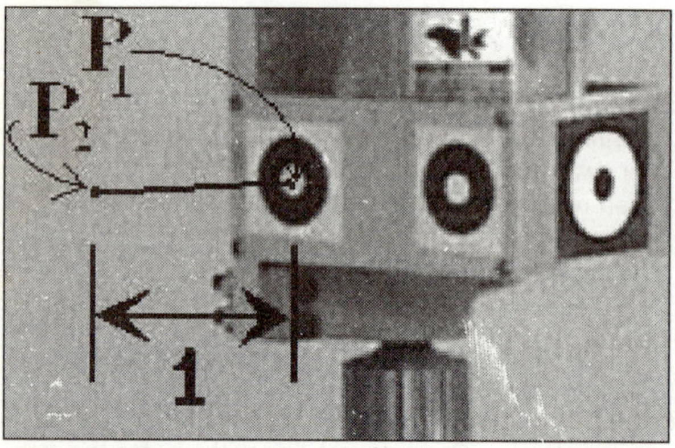

Figure 7-19. The nominal forward kinematics model can be applied to produce an adequate approximation to the unit normal of a given nonunique cue referred to the robot-base-fixed reference frame.

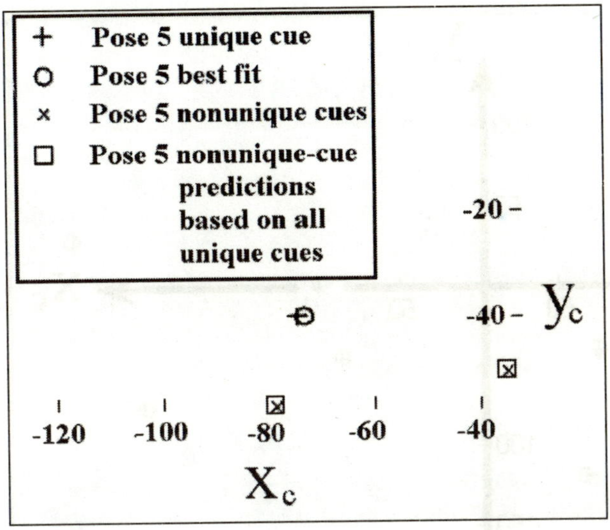

Figure 7-20. Magnification of fifth-position best fit of the unique cue, combined with fifth-pose detections of nonunique cues and predictions of these locations.

Equal weighting with such a poor global model does not always produce such great confidence in the ability to distinguish one nonunique cue from another based upon proximity of detected camera-space location to prediction. The situation can be improved however with nonuniform weighting and application of J as defined in Eq. 7.16 above. For example, if the following simple formulation for $\mathbf{W}_i$, i=1,2, ... , n, is used, the predictions and fits of *Figure 7-20* ordinarily improve: $\mathbf{W}_i = 1/[(5-i)^2+1]$. With close predictions, a simple computer logic to identify the nonunique cue indications will work reliably.

Three general points concerning this particular procedure for cue identification: ***First***, if the procedure for distinguishing among nonunique cues based upon proximity does not yield a clear result, the data should be dropped. Wrong cue identities at any stage need to be avoided as they invalidate the entire procedure. ***Second***, a weighting scheme such as $\mathbf{W}_i = 1/[(5-i)^2+1]$ above depends upon a particular strategy for defining the preplanned poses: contiguous movement. Although taken as a whole the poses should represent a wide swath of joint space, camera space and physical space, it is best to make the movement between consecutive entries in **Table 7-3** joint rotations comparatively small. ***Finally***, if the cameras are very near to the robot it may be that even the skewed weighting above will not yield the confidence needed to identify nonunique cues. For such a case the flattening procedure mentioned above (see *Figure 7-21*) can be utilized. In particular, it is possible to flatten the unique-cue data about a predicted nonunique-cue camera-space position and compare this prediction against actual indications for the pose in question. This will of course require the simultaneous consideration of the other cameras.

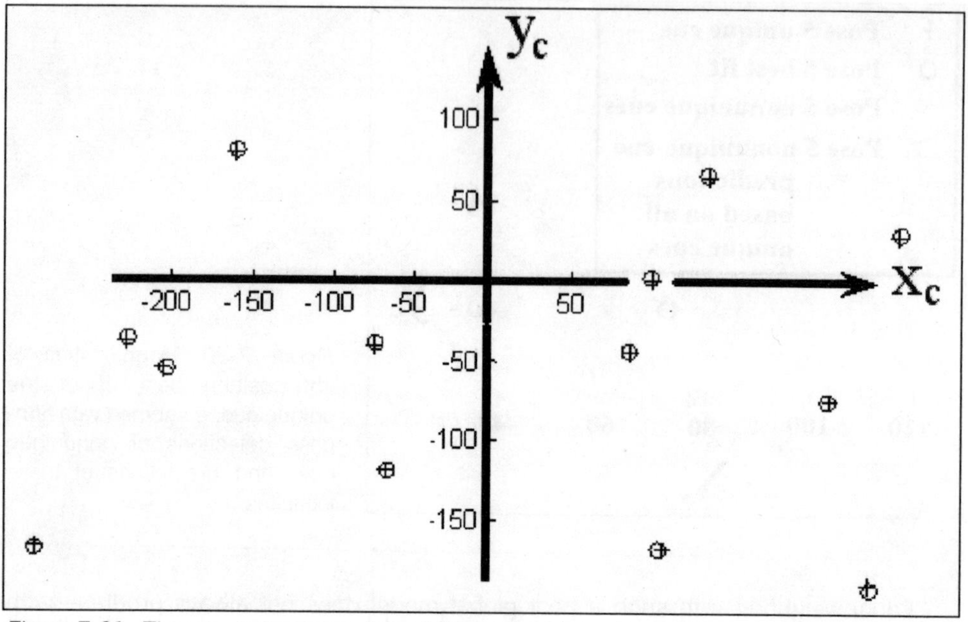

Figure 7-21. The raw, unique-cue data of *Figure 7-11* have been flattened - in this case about the fifth unique-cue position of **Table 7-3**. The dramatic improvement in best fit is an indication that the predominant source of model error when using the orthographic camera model can be eliminated through flattening.

Any maneuver can begin with $\underline{\mathbf{C}}$ as initialized above. Applying an algorithm similar to that of *Figure 7-13*, new data can be included and given preferred weight in the estimation of elements of $\underline{\mathbf{C}}$ in accordance with:

$$J(\underline{\mathbf{C}}) = \Sigma_i \Sigma_j \; W_{ij} \{[x_{cj}{}^i - f_{xj}(\underline{\boldsymbol{\theta}}^i;\underline{\mathbf{C}})]^2 + [y_{cj}{}^i - f_{yj}(\underline{\boldsymbol{\theta}}^i;\underline{\mathbf{C}})]^2\} \quad (7.28)$$

where the double summation indicates summation (i) over all poses where end-member cues have been identified, and summation (j) over all end-member cues that were detected in the ith pose. The functions f_{xj} and f_{yj} are defined according to

$$\begin{aligned} f_{xj}(\underline{\boldsymbol{\theta}}^i;\underline{\mathbf{C}}) &= (C_1^2+C_2^2-C_3^2-C_4^2)\, g_x^*(\theta_1, \theta_2 \ldots, \theta_n; \mathbf{X}_j, \mathbf{Y}_j, \mathbf{Z}_j) + 2(C_2C_3+C_1C_4) \\ &\quad g_y^*(\theta_1, \theta_2 \ldots, \theta_n; \mathbf{X}_j, \mathbf{Y}_j, \mathbf{Z}_j) + 2(C_2C_4-C_1C_3)\, g_z^*(\theta_1, \theta_2 \ldots, \theta_n; \mathbf{X}_j, \mathbf{Y}_j, \mathbf{Z}_j) + C_5 \end{aligned}$$

(7.29)

$$\begin{aligned} f_{yj}(\underline{\boldsymbol{\theta}}^i;\underline{\mathbf{C}}) &= 2(C_2C_3-C_1C_4)\, g_x^*(\theta_1, \theta_2 \ldots, \theta_n; \mathbf{X}_j, \mathbf{Y}_j, \mathbf{Z}_j) + (C_1^2-C_2^2+C_3^2-C_4^2) \\ &\quad g_y^*(\theta_1, \theta_2 \ldots, \theta_n; \mathbf{X}_j, \mathbf{Y}_j, \mathbf{Z}_j) + 2(C_3C_4+C_1C_2)\, g_z^*(\theta_1, \theta_2 \ldots, \theta_n; \mathbf{X}_j, \mathbf{Y}_j, \mathbf{Z}_j) + C_6 \end{aligned}$$

where $\mathbf{X}_j$, $\mathbf{Y}_j$, $\mathbf{Z}_j$ are the fixed end-member-based coordinates of the jth cue.

This continuous-improvement process should ensure good camera-space kinematics estimates for the positioned point $\mathbf{P}$,

$$f_{xP}(\underline{\boldsymbol{\theta}}^i;\underline{\mathbf{C}}) = (C_1^2+C_2^2-C_3^2-C_4^2)\, g_x^*(\theta_1, \theta_2 \ldots, \theta_n; \mathbf{X}_P, \mathbf{Y}_P, \mathbf{Z}_P) + 2(C_2C_3+C_1C_4)\, g_y^*(\theta_1, \theta_2 \ldots, \theta_n; \mathbf{X}_P, \mathbf{Y}_P, \mathbf{Z}_P) + 2(C_2C_4-C_1C_3)\, g_z^*(\theta_1, \theta_2 \ldots, \theta_n; \mathbf{X}_P, \mathbf{Y}_P, \mathbf{Z}_P) + C_5$$

(7.30)

$$f_{yP}(\underline{\boldsymbol{\theta}}^i;\underline{\mathbf{C}}) = 2(C_2C_3-C_1C_4)\, g_x^*(\theta_1, \theta_2 \ldots, \theta_n; \mathbf{X}_P, \mathbf{Y}_P, \mathbf{Z}_P) + (C_1^2-C_2^2+C_3^2-C_4^2)\, g_y^*(\theta_1, \theta_2 \ldots, \theta_n; \mathbf{X}_P, \mathbf{Y}_P, \mathbf{Z}_P) + 2(C_3C_4+C_1C_2)\, g_z^*(\theta_1, \theta_2 \ldots, \theta_n; \mathbf{X}_P, \mathbf{Y}_P, \mathbf{Z}_P) + C_6,$$

estimates that are highly accurate in the vicinity of the terminus. Writing a real-time program to coordinate this movement/estimate-improvement synchronization is a challenging task, particularly with the always-wise flattening of all camera-space samples about the target surface point. It is a good challenge for a good programmer even if the robot is commanded to pause at junctures where new images are acquired for

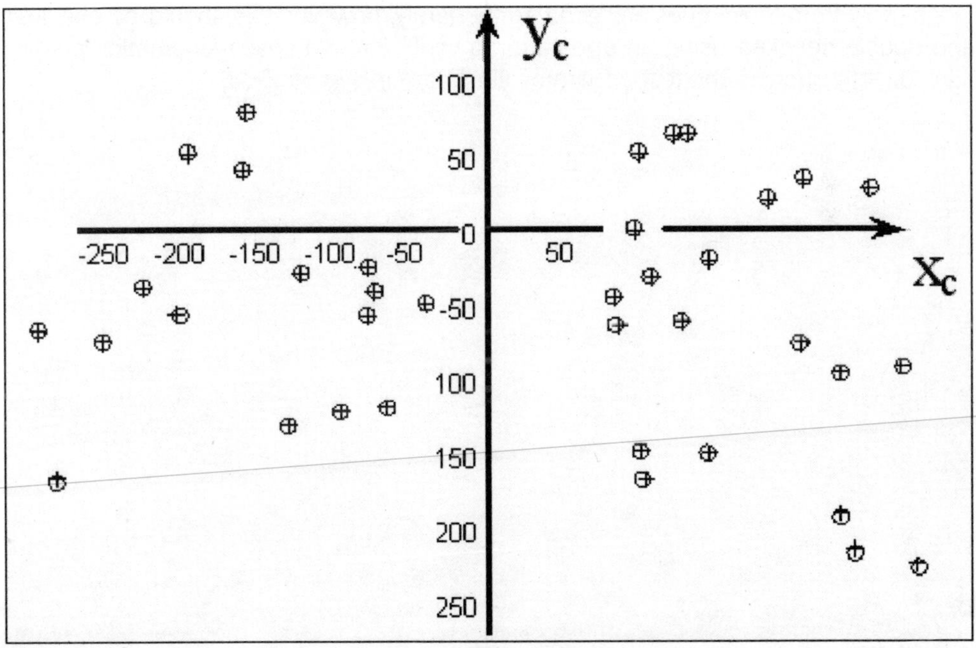

Figure 7-22. The excellent fit across a broad range of joint space of the flattened Table-7-3 data from the unique cue as well as nonunique cues implies two things: First, the skewness of preferred weighting favoring the vicinity of the terminus should be *gradual*; and second, prospects for using cues to estimate the camera-space kinematics of end-member junctures some distance away from those cues are good.

upgrading **C** estimates. But making the whole thing operate continuously, without pauses to acquire the new images en route, is more challenging still. Yet, with today's fast computers and the intrinsic ability of the present approach to tolerate finite time delay for image/estimation processing – queuing robot reference joint poses systematically slightly ahead of the place of current execution - it is possible to achieve.

The inherent redundancy of incoming information not only results in statistical precision and certainty but also can be used by a good programmer to reject or accept with a high level of certainty any incoming data from the image analyzer. *Figure 7-22* shows visually the high level of consistency that attends incoming data correctly identified and correctly processed. It shows both the flattened unique-cue data and the flattened nonunique-cue data following identification of all indications in **Table 7-3**.

The consistency of the high amount of flattened data of *Figure 7-22* with the six-parameter model carries with it two other messages – about weighting, and about the use of end-member cues to identify camera-space kinematics of a point somewhat distant from these cues. First, the skewness of preferred weighting favoring the vicinity of the terminus should be *gradual*; and second, prospects for using end-member cues to estimate the camera-space kinematics of end-member junctures some distance away from those cues are *good*. Importantly, however, this should be checked and double checked using an apparatus to verify camera-space-kinematics predictions directly such as the test apparatus illustrated in *Figure 7-14*.

CHAPTER 8

POINT-AND-CLICK HUMAN SUPERVISION—WHAT PEOPLE DO WELL

When thinking of robots doing all or a portion of a job now done by humans, it is tempting to construe tasks as traditionally sequenced and performed. Yet such a direct transfer of task execution is neither necessary nor desirable since much of the evolved art of human task execution is a specific accommodation to human limitations not likely to be shared by the robot. Machines have tremendous ability when guided by remote cameras - cameras that are separated each with its own point of view and consequent geometric advantage. Much more than any human they can deliver a tool with precision, steadiness, directed force and reliability. But it will often be up to the human supervisor to specify "where", "at what orientation", "how", and/or "how much". Such specification of course must be done in a way that both human and machine understand.

The discussion of Chapter 6 introduces Camera-Space Manipulation with the imagined task of manipulating a felt pen tip to draw a diagram onto an arbitrarily located ostrich egg. The discussion only advances this process to the point of initial contact. The present chapter develops the methods from this starting point, expanding to a very broad range of practical tasks. We begin with the drilling task of *Figure 8-1*.

Drilling a hole.

Hole drilling in this context is a five-degree-of-freedom task. That is, in general it will require five separate joint rotations, for example θ_1 through θ_5 of the six-axis robot diagramed in *Figure 2-1*. In the case of the arbitrarily placed log of *Figure 8-1*, we desire to drill exactly one cm deep at a juncture and angle relative to a plane as prescribed by the human user or supervisor. In this illustration the angle of entry relative to the user-designated plane is ninety degrees. These aspects of human task specification are achieved using point and click followed by autonomous direction of a laser beam of light as indicated in Chapter 6 for pen-tip touch-down onto the ostrich egg.

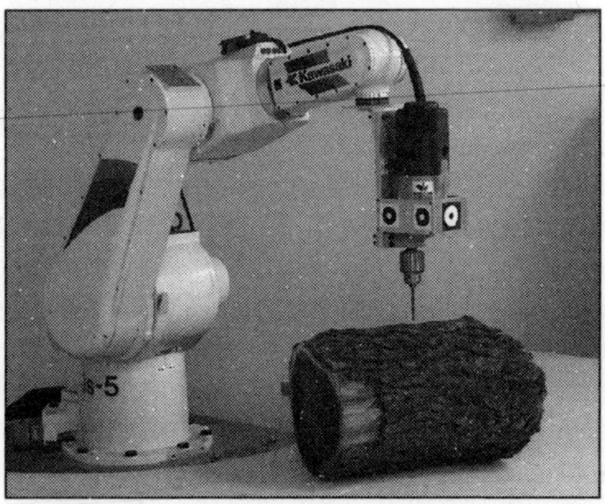

Figure 8-1. Kawasaki Js5 robot. At least five of the six axes of the Kawasaki Js5 robot are needed to drill a hole of prescribed depth normal to the surface at a user-selected point.

As per the Chapter 6 discussion, and illustrated in *Figure 8-2*, user involvement entails a mouse click onto an image of the workpiece acquired from a conveniently placed, often user-directed camera, called herein the "selection camera". This procedure designates the target point, the location where the robot is to drill the hole. The user also specifies the orientation of the entering drill bit by pointing and clicking onto three additional junctures on the selection camera's image of the surface of interest. These three points on the physical surface designate a plane perpendicular to which the hole must be drilled - similar to the discussion in Chapter 6. A useful aspect of this is that human knowledge and intuition regarding the geometric characteristics of the surrounding surfaces as well as the geometry of the log can be applied in a very straightforward way. If, for example, as indicated in *Figure 8-3*, the log in question is attached to a board on one end, the user can recognize this and avoid camera-space targets for the laser pointer by clicking onto log-surface points that are near to the entering location but not located on irrelevant surfaces. Human judgment and physical interpretation of portions of the image presented on the monitor must be used, but for most useful applications this judgment is good.

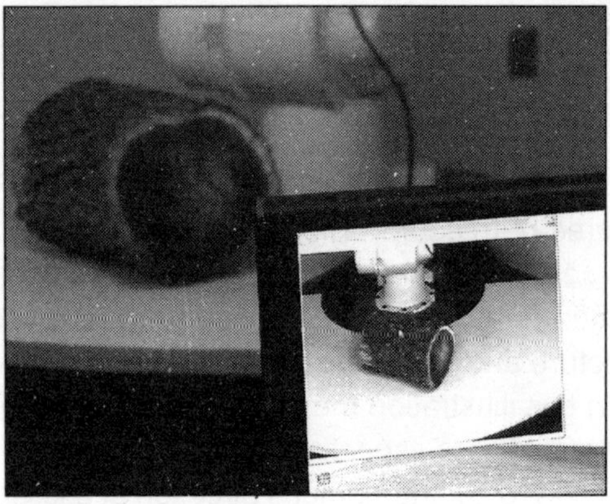

Figure 8-2. A human supervisor (who may be very remote from the physical system) points and clicks to select point of drill entry. The monitor is shown in foreground; the actual log is in background.

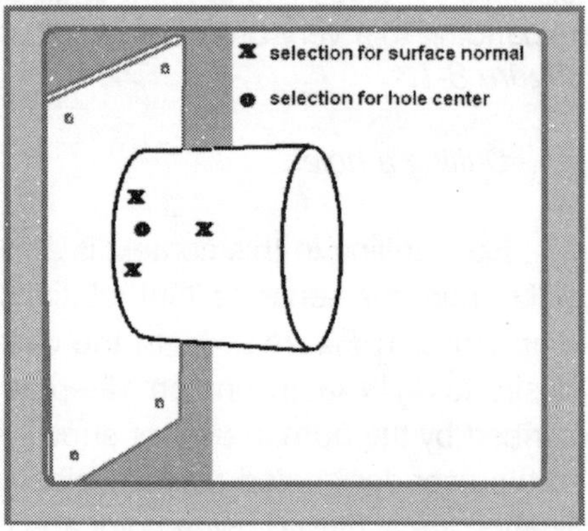

Figure 8-3. Monitor used by human supervisor. After selecting the surface location for drilling, three other points are selected judiciously to define the plane perpendicular to desired drill orientation.

The convergence of laser spots, and *everything* else leading up to and including the physical drilling occurs fully autonomously. Laser spot convergence is used for collocating the center of a laser spot onto all four of the user-selected points of *Figure 8-3*, one convergence event at a time. A conveniently located pan/tilt-unit-borne laser pointer is used together with an approximate Jacobian, determined and used as discussed in Appendix D. The selection camera, shown in *Figure 8-4*, provides the reference frame within which the spot convergence is directed. It is interesting that essentially three-dimensional information is provided (via spot detection in the other,

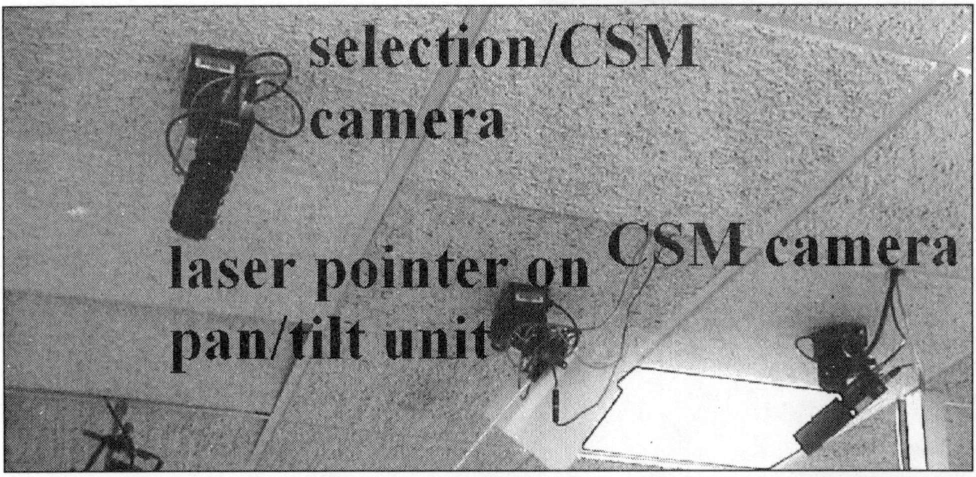

Figure 8-4. A configuration which includes arbitrarily placed selection camera and CSM cameras, as well as laser-bearing pan/tilt unit. Note that the selection camera may be designated as any of the cameras, and may or may not for a given maneuver also serve as a CSM camera. A minimum of two cameras, total, must serve as CSM cameras.

non-selection or "CSM" cameras which also are shown in *Figure 8-4*) using control in a two-dimensional reference frame. The reason this is possible is because all physical surfaces in all participant images, including for instance the board and wall and cylindrical-log surfaces of *Figure 8-3*, represent constraints that lower the dimensionality of the pertinent portion of physical space from three to two. So object surfaces make mapping from physical space into camera space 2D to 2D much as the actual contours of hills and valleys make mapping of land onto a piece of paper a 2D-to-2D exercise. Of course depending upon the positions and/or orientations of participant system components, including cameras, target log surface, board (*Figure 8-3*), and laser/pan-tilt-unit, there may be important physical-surface points whose locations in selection-camera space, or one or more of the other CSM-camera spaces, cannot be identified via laser convergence. If this is true for the selection camera, then the user must make a provision such as shifting to a different selection camera, or pan/tilting remotely the existing selection camera, to correct the matter. If it is true for one or more CSM cameras then the versatility of CSM to transition control from one group to a different group of cameras can be used. Moreover, this decision can be programmed and executed automatically at the remote site.

Convergence of the laser pointer onto the initial juncture, as selected in *Figure 8-2*, is ordinarily rapid, but can be comprised of an irregular sequence of via points, as indicated in *Figure 8-5*. Factoring in to this sequence is the actual geometry of the intermediate objects upon which the laser spot falls as well as the geometric relationship between selection camera and pan/tilt unit. Particularly if these two are not physically close one to another, it becomes more likely that, at one or more junctures of the

Figure 8-5. Sequence of laser spots falling on intermediate surface points en route to the user-selected target.

laser spot en route to its designated terminus, physical objects will cause the spot to be hidden from the selection camera. Continuation of convergence, however, can normally be advanced in such a situation by a program logic that commands small but arbitrary motion of the pan/tilt unit as required to reestablish visibility of the laser spot to the selection camera.

Figure 8-6. The drill is posed at its first stop en route to hole completion. Several intermediate points **P** are added to the first and last, which are both shown here, in order to assure gradual entry into the log.

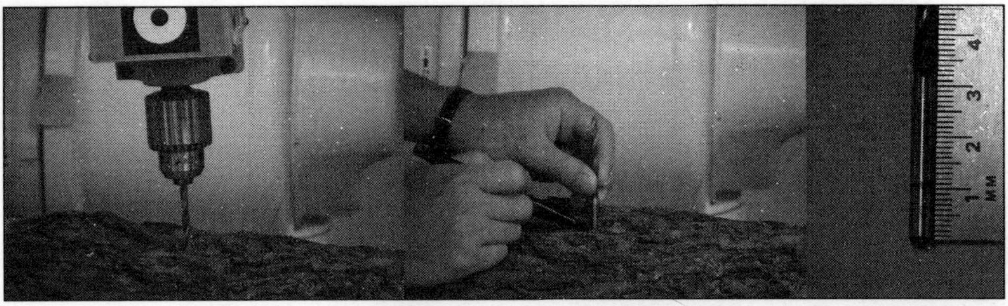

Figure 8-7. The robot delivers drill bit to its final depth. The back of the drill bit inserted into the drilled hole is marked to indicate that actual hole depth is as desired.

Following detection of all four laser spots, converged to a target indicated for instance in *Figure 8-2,* the robot moves to create the drilling event. The initial robot-execution pose can and should occur with the drill-bit tip some distance from the surface, as indicated in *Figure 8-6.* In accordance with Chapter 6, this is done by defining the point **P** that is to be collocated with the physical entry point to be along the geometric extension of the bit, in the case of *Figure 8-6* fully five centimeters beyond the physical tip juncture. Orientation control here and throughout subsequent entry of the drill into the log is commanded in accordance with the algorithm outlined in Chapter 6. (See *Figure 6-14.*) Both position control and orientation control are refined during approach. They make use of cues, shown located on the end member in *Figure 8-1,* to estimate and refine during closure the camera-space kinematics in all participant CSM cameras, as discussed in Chapter 7.

Transition of the defined end-member point P that is to be collocated with the selected surface-entry point along the drill bit occurs gradually, terminating 1cm from the drill-bit tip as indicated in *Figure 8-6. Figure 8-7* shows the drill in this final physical position; the consequent hole is 1 cm deep.

Orientation control

As discussed in Chapter 6, axes 4 and 5 of the robot affect orientation, more than position of point **P**; and axes 1, 2 and 3 affect position of point **P** more than orientation. However, in general there is coupling between the movement of any of the five axes of rotation and position/orientation. Chapter 6 goes on to indicate that nonetheless an iterative procedure to solve for all five joint angles will converge efficiently if position objectives are corrected at each iteration using θ_1, θ_2 and θ_3, while orientation objectives are corrected using θ_4 and θ_5.

The latter correction is based upon the unit normal to the surface, or rather an approximation to that unit normal, determined with user-selected surface junctures as indicated in *Figure 8-8.* In that figure the selection-camera image, as presented to the

Figure 8-8. The user designates both the location o of the hole and the drill-bit orientation by pointing and clicking onto the four junctures shown.

user, is shown. Each of the four "+" marks represents a juncture identified via GUI by the user. Item o is the target point of drill entry and items A, B, and C are the junctures used to define the surface's unit normal at the point of drill entry. They correspond with points A, B, and C in *Figure 6-13*, and the unit normal is computed in accordance with the associated Chapter 6 discussion - with the modification based on Chapter 7 discussion of *flattening*. Points A, B, C and o are all located via laser-pointer convergence in all participant camera spaces.

Figure 6-13 shows a flow chart used in the iterative process of achieving five-degree-of-freedom position/orientation control for perpendicular delivery of a pen tip onto an arbitrary surface. This same approach is applicable to the present problem of drill-bit delivery. One modification, however, that is computationally cheap and very much worth including relates to Chapter 7 flattening procedure for position control as indicated in the flow chart of *Figure 7-10*. This flattening was used to create the drilling maneuver, based upon the selection of *Figure 8-8*, whose full-depth terminus is pictured in *Figure 8-9*.

However, in addition to the application of flattening to point-P positioning, flattening was also used for purposes of improving orientation control using points A, B, and C of *Figure 8-8*. The flattening process outlined in Chapter 7 has an improving effect on the assessed orientation estimates of each camera with respect to the reference frame of the action of the drill. This improvement should be used to improve the drill's actual orientation relative to the plane of points A, B and C. Note that Chapter 7 algorithm for orientation control entails calculation of nominal physical position of these three points. The improvement of flattening is realized by merely applying the flattening procedure of Chapter 7 not only to identify the nominal physical location of target point o but also the nominal physical (x-y-z) locations of A, B, and C, prior to taking the cross product of *Figure 6-13*.

Figure 8-9. Although the hole created lies in a different surface location compared with that of *Figure 8-6*, the two poses correspond with tool-fixed locations of point **P** at the two junctures defined in *Figure 8-6*. Intermediate locations for point **P** define the interim drill bit location relative to the log. Orientation of the drill is based on *Figure 8-8* A, B, and C.

More generality

Consider the cutting task illustrated in *Figure 8-10*. Two end points of an incision are selected by a human user, points A and B. The surface is continuous and smooth but arbitrarily curved. The desired motion is an in-and-out cutting action.

Figure 8-11 illustrates one aspect of the steps needed to realize this motion using camera-space manipulation - the sequence of points **P** defined relative to the coordinate system **X Y Z** of the end effector. $\mathbf{P}_1$, $\mathbf{P}_2$, $\mathbf{P}_3$ and $\mathbf{P}_4$ correspond to the four instants indicated in *Figure 8-10*. $\mathbf{P}_3$ and $\mathbf{P}_4$ indicate the reciprocating sawing action of the blade. This can be repeated as many times as desired over the course of the cut merely by adjusting the number of cycles of redefinition between $\mathbf{P}_3$ and $\mathbf{P}_4$ of point **P**. In particular, the target point for **P** will move gradually from user-defined point A to user-defined point B. As "s" of *Figure 8-12* increases from **0** to s_{max} a large number, perhaps 100, changes in the **X-Y-Z** definition of **P** will be supplied. That is where the opportunity comes to define the amplitude and number of cycles of the reciprocating movement.

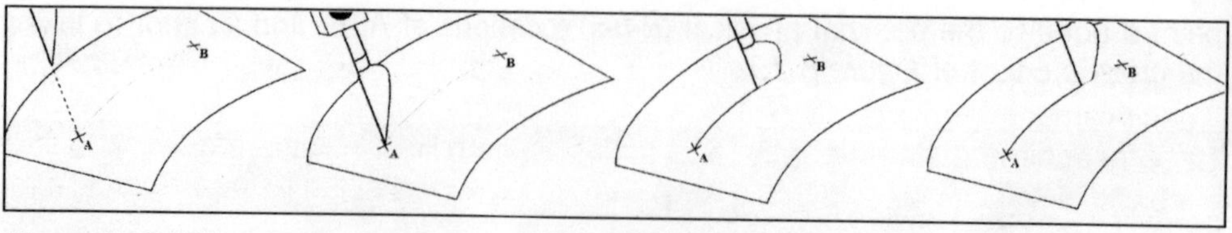

Figure 8-10. The user designates end points for making a cut, and up-and-down end-effector "sawing" motion ensues based upon sequential shifting of P along the blade edge.

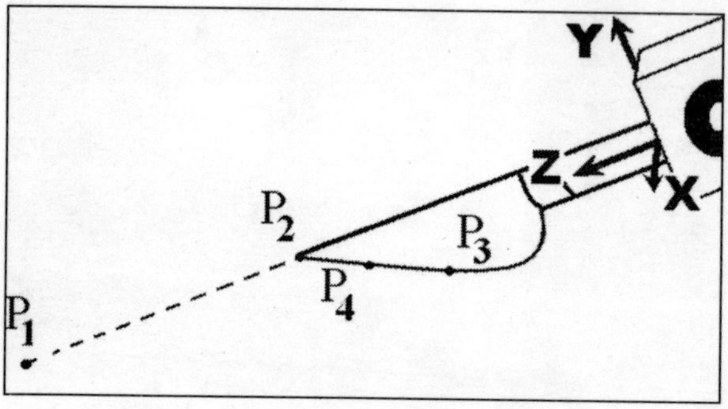

Figure 8-11. Perhaps 100 sequential locations **X Y Z** of **P** will be defined for the slicing or sawing motion of *Figure 8-10.* The four instants shown in *Figure 8-10* correspond, respectively, to $\mathbf{P_1}$ through $\mathbf{P_4}$ above.

In order to relate the sequence of defined positions **P** to distance along the incision s, it is necessary to identify in each participating camera space the locus of points connecting user-selected end points A and B. Of course these points must lie on the surface of interest, but even with that constraint there is an infinite number of possibilities. One possibility would be a straight-line connection in the selection-camera space (which generally will not, depending upon the surfaces and camera, be a straight line in physical space). This is the easiest surface-point interpolation to define. Another possibility would be the surface-contour line where the physical length s_{max} is minimum.

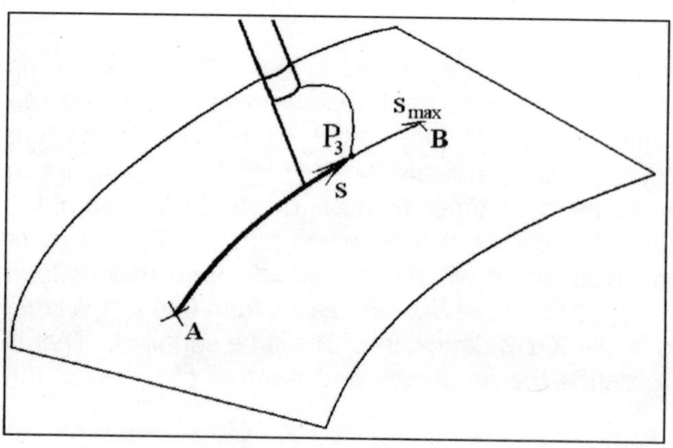

Figure 8-12. $\mathbf{P_3}$ is placed as shown for the particular fraction s/s_{max} of incision completion indicated in the diagram.

Either of these alternatives, indeed any strategy for establishing a surface-contour trajectory between A and B, requires a continuum of points – with point correspondence among cameras - identified in all participant CSM cameras. One efficient strategy for achieving this entails use of a multiple-beam laser pointer. *Figure 8-13* shows a piece of soft clay formed into a wavy shape. As shown in the figure, directing a multiple-beam laser pointer onto this surface leaves a kind of matrix of laser spots.

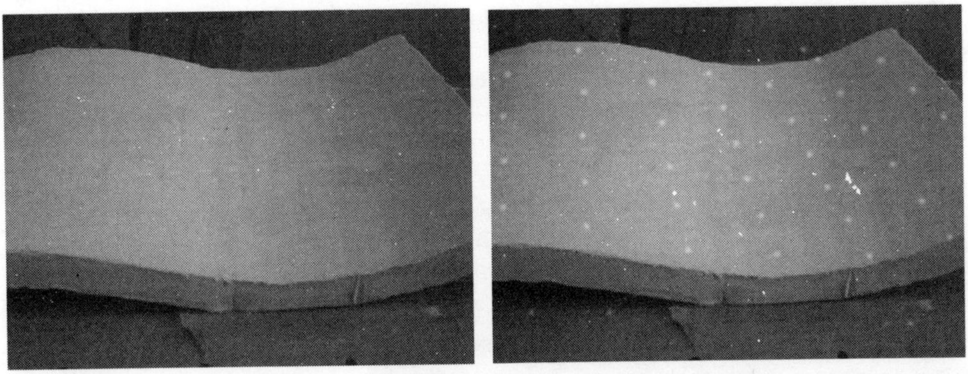

Figure 8-13. Laser spots falling on the surface as shown in the right can be identified in each of the participant camera spaces via image differencing of the case pictured on the left from that shown on the right, as discussed in Appendix D.

Image differencing results in an ability to locate most of the surface laser spots in all participant CSM cameras. If these spot indications are matched among CSM cameras, then the procedure of flattening about each matched spot allows for the nominal three-dimensional location of the laser spots. Of course assessed three-dimensional coordinates of the spots will not correspond with any known reference frame (though nominally it is a frame fixed with respect to the robot base), but if the weighted $\underline{\mathbf{C}}$ estimates of each camera are used in the same way for all spots, that frame will be consistent with local actual robot-kinematics. (See the Chapter 7 discussion concerning nominal vs actual robot kinematics.) This spot location allows for accurate positioning of the manipulated juncture **P** using the known, nominal kinematics model. Chapter 7 discusses the equivalence between joint-rotation robot-pose specification and specification of robot-pose in terms of nominal physical coordinates. The nominal robot-kinematics model (Appendix B) discusses this correspondence. In the present instance a point **P** located at the desired inscription depth from the tip of a rigid wire is identified to be collocated with a moving target point along the surface of the clay, in order to realize the UASLP inscription.

Spot matching among participant cameras is facilitated by the relatively wide separation among spots as indicated in *Figure 8-13*. A prior autonomous positioning (*Figures 8-5* and *8-8*; also Appendix D) of three single laser spots on the user-selected region - closely spaced relative to the surface curvature - enables computer execution of a matching logic based upon a near-linear mapping within that small region among cameras. (For surfaces with small radius of curvature a scheme more locally elaborate than simple linear mapping/correspondence among cameras may be needed.) Importantly and helpfully, the pan/tilt unit that was set at one position for the right-hand-side image of *Figure 8-13* can be moved, slightly and autonomously, to allow for another, similar image – but one with spots slightly displaced from their shown position. As there is no danger of mismatching spots from different images this procedure can be continued to result in an almost arbitrarily dense group of spots. That in turn permits a logic for migrating the spot-matching region away from the original single laser spots to cover any size region on a continuous surface. These many matched spots allow for the aforementioned ***x-y-z*** surface-point coordinates to be computed with respect, nominally though not actually, to the robot's physical frame.

A least-squares surface-contour fitting can be accomplished using the densely packed nominal physical coordinates of the locus of points connecting A with B of *Figure 8-12*, or, in the instance of the inscription of *Figure 8-14*, the complex UASLP acronym.

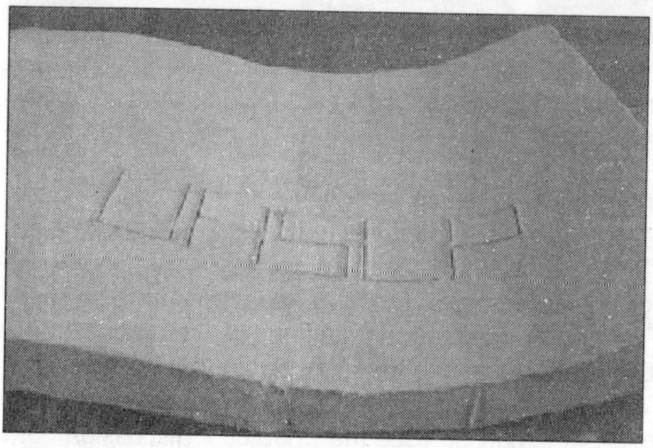

Figure 8-14. Dr. Emilio Gonzalez's group at Mexico's Universidad Autónoma de San Luis Potosi uses multiple, matched laser spots together with CSM to inscribe University initials at a prescribed depth into an arbitrarily contoured soft clay using a rigid wire at the end of the robot tool.

The actual mapping of UASLP from the flat template onto the arbitrarily curved surface makes use of three rules: First, corner junctures, where one straight line meets another on the flat template, translate to same-angle direction changes relative to the tangent plane of the actual surface at the point of the turn. This is like the idea of a conformal map. Second, the arc length across the physical surface is the same as that of the template. And third, the length of the line from its particular starting point to its particular terminus is the minimum for the surface in question (provided that surface is of the common "developable" type.)

Because the clay blank of *Figure 8-13* entails surface curvature about a set of parallel axes only, it becomes possible to ensure the accuracy of the UASLP shape by conforming a flexible plastic sheet with the inscription. As shown in *Figure 8-15* alignment between the two is perfect. This is an important observation. It means that the versatility of the pen plotter of *Figure 6-1* to transfer any mathematical form to a flat sheet of paper has been brought to the world of general surfaces. More than that, the *calibration* inherent to the 2D pen plotter has no counterpart with this technology. Neither cameras nor robot are calibrated. Let a human user specify the shape, depth, starting point, scale, and starting orientation, and the uncalibrated system takes over from there. The result is virtually perfect. This same facility can be brought to countless types of operations. For example, *Figure 8-16* shows the same UASLP contour in the form of a weldment. The inherent ability of the present system to enforce a constant tip speed becomes particularly useful for this class of application.

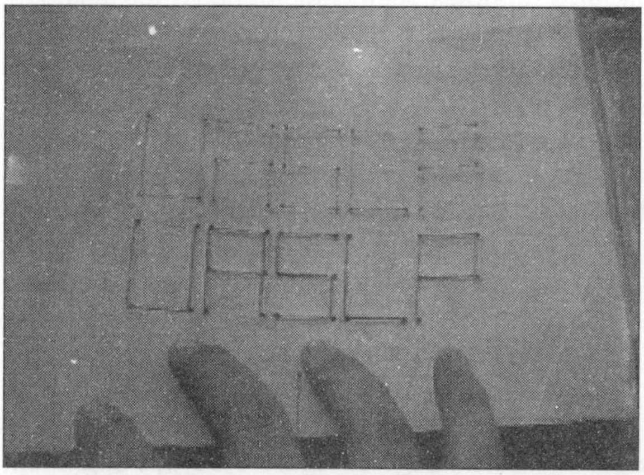

Figure 8-15. UASLP printed to size on plastic bends to exact fit.

Figure 8-16. The same UASLP of Figure 8-13 is welded - this time onto a flat metal surface.

As mentioned above, the particular algorithm used in *Figures 8-15* and *8-16* to transfer straight lines from the template onto an arbitrarily curved surface, that of the geodesic curve, is just one of many possibilities. This particular choice has the property that, if the template's trace closes back onto itself, the transferred shape will not generally meet at the same point on the arbitrary surface. For this and other reasons, task specification such as welding may, more commonly, entail human specification of several surface "*via*" points between which some kind of desirable interpolation strategy is defined. There is a great deal of flexibility in this regard. The incision task of *Figures 8-10*, *8-11*, and *8-12* differs from the inscription task of *Figures 8-14* and *8-15* in that it connects two specific points. Users might specify any number of such surface points using point and click for a complex seam-welding task, for instance. Any of several interpolation strategies, including least-distance strategies, could be invoked for actual welding.

One additional difference between the inscription of *Figure 8-14* and the incision task of *Figure 8-10* is that the latter requires an additional degree of robot freedom – for a total of six – compared with the UASLP clay inscription, which, like the ostrich-egg or drilling problem, requires just five. The reason for this additional need relates to the requirement of holding the cutting-tool blade in such a way that it is aligned with the direction of the incision. In this sense, the incision task represents the most general type of six-degree-of-freedom, rigid-body positioning of one three-dimensional object relative to another.

At the particular position or pose of the third image of *Figure 8-10* this alignment requirement can be described as follows: The surface-parallel unit vector **e** of *Figure 8-17* must be pointed in a direction opposite to the tool's **Y** axis as shown in *Figure 8-11*. This shifting orientation requirement (shifting as s of Figure 8-12 advances) is added to the also-shifting perpendicularity requirement that *Figure 8-15* **Z** axis must oppose the current-s (*Figure 8-12*) surface normal **n**. Both **n** and **e** can be found from the of the locus of ***xyz*** coordinates of the trace from A to B. Modification of the procedure of the block diagram of Chapter 6 to accommodate this additional specification is straightforward.

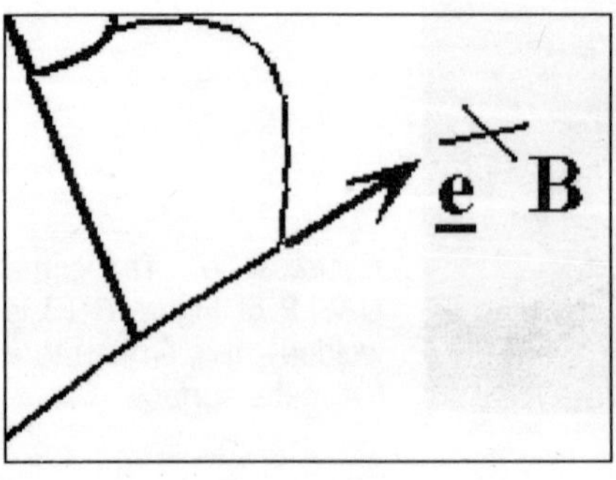

Figure 8-17. The unit vector **e** remains in the plane of the surface but moves within that plane as the point of edge contact s advances.

Moreover

The above examples portend a broad range of prospective uses. Provided surfaces in question remain fixed relative to cameras that register and match among one another points or junctures highlighted by laser spots, the defined action of the robot's end effector is reliable. It can be made by judicious choice of **P**-location movement and orientation requirement to vary to include many useful tasks. Freedom from calibration, and zero-mean, low-variance expectation of local camera-space-kinematics estimates, together with facility of adding more cameras to the resolution of joint rotations means cheap, robust ways to improve maneuver precision. The ability to compute camera-space-kinematics refinements from delayed image information pertaining to end-member surface points that are visually accessible - even as the maneuver ensues - means ready and convenient real-world implementation.

All kinds of surface operations including scrubbing, painting, sanding, deburring, and scanning are straightforward extensions of the above. A user might prescribe a region on a surface to be exhaustively covered by the operation using point and click. From there, CSM-directed surface action can be computed and refined in real time using, for example, sweep action where separation between passes is a fixed, user-input distance.

Assembly, comprised of picking up an object and positioning it precisely relative to a second, stationary body, is similarly possible to achieve. Nothing prevents the user from panning, tilting or zooming the selection camera to achieve an ideal view from which to precisely specify object junctures relative to which either of these two operations - picking up or positioning – is to occur. As with drilling, specification of the geometry of the immediate approach of the manipulated object with respect to the target object is a matter of previously, judiciously defining a sequence of positions of point **P** as with the drilling task in *Figure 8-6*. If the geometry with which a grasped object is oriented relative to cues on the end member must be refined, it is possible for the user to command the robot to position its end member precisely relative to a selection camera. The user's point-and-click instruction regarding image location of key object junctures can, combined with CSM, ensure highly precise characterization of the six coordinates that describe the relative rigid-body juxtaposition of object relative to grasper. This in turn allows grasper-fixed cues to be used with CSM for guidance of the engaged object onto, or into, a target second object.

Machining, drilling, cutting, carving, rasping, planing, and other kinds of material-removal action can be achieved with high precision. Even sculpting a complex three-dimensional shape from a wood, stone or metal blank is possible. Shaving layers of material as the detected surface approaches its reference (or desired) contour is straightforward with a repeated application of the laser-spot treatment described above. And of course the addition of material as with welding or a fused deposition to build a solid prototype is merely the reverse of this operation. With either the material-adding or material-removing tasks there will be some "drift" - some departure from the 3D template stored in software - as the extent of the created surfaces increases beyond the user-initialized junctures. With the UASLP inscription of *Figure 8-12*, for instance, although depth of inscription, which refers directly to the as-located surface, will remain accurate indefinitely, departure of the in-plane shape from the template will begin slowly to grow – even though camera-space kinematics are kept current with the current joint pose A – with inches from the user-initialized juncture. This effect can be arrested however with previous markings across the region of interest in the uncalibrated cameras' reference frames. For instance, if the task is to replicate Michelangelo's David from a geometrically square and plum block of stone, participant cameras could register prior to material removal several laser-spot-highlighted junctures at precise intervals on the initial block. These in turn serve as a kind of guide to prevent the aforementioned drift during material removal. Camera-control authority can be shifted gradually among groups of participant CSM cameras by introducing in software a logic to adjust gradually the weighting over camera index as vantage-point advantages shift during execution - from full weighting in the resolution of joint pose to zero weighting. Similarly, the weight given to any one of the above-mentioned camera-registered initial-block surface points can be shifted - increased or decreased all the way down to zero – again, automatically, in software, with the spot's proximity to the region currently being sculpted. The interesting idea is that the constraints that give rise ultimately to the desired surface contour are not mechanical – calibrated kinematics, known grinding-surface wear, and the like – but rather optical, in the form of relatively distant out-of-the-way cameras' maneuver requirements, cameras whose presence and mapping of physical space into camera space does not shift during operation. (An explanation concerning robustness of the above to grinding-surface wear is in order. As indicated in the beginning of Chapter 7, CSM is indeed predicated on the assumption that the end effector is a known rigid body, in particular that the tool-relative geometric relationship between cues and positioned juncture is known and does not shift, as with tool wear. Depth control of the inscription of UASLP, for example, depends upon the length of the inscribing wire being known and constant. With removal of material in layers, however, and the modus operandi of re-evaluating the uncovered layer using laser spot appearances in camera space, this grinding-surface wear can be inferred accurately and automatically. Thus the known adjustment in tool-contact points' position with respect to the cues is adjusted.)

One strategy for duplicating surfaces, either by depositing or removing material, avoids difficulties with drift directly. This consists of placing the prototype shape, the shape that is to be duplicated, into the fields of view of participant cameras, and registering/matching large numbers of laser spots, densely packed, in the camera spaces. The prototype is then removed and replaced by the blank whereupon the CSM operation of removing or adding material ensues. Provided control-camera transition is gradual, as per the above, entirely distinct groups of cameras can be responsible for the various regions of the recreated object. The requirement is that camera mappings of physical space into camera space remain constant and unchanged throughout the operation.

Importantly, all of this capability is **calibration-free**. Everything is done by correcting action, as with a human being, in the reference frames of the visual sensors. Unlike the human, however, hand/tool action relative to an object can be realized in three dimensions with quantitative specificity.

CHAPTER 9

HOLONOMIC VS NONHOLONOMIC ROBOT KINEMATICS

An essential difference separates the way these two types of robot can be controlled.

Both holonomic and nonholonomic systems are part of our everyday world experience. They are so familiar and our ability to control them so accomplished that it becomes hard to discuss the related issues: Our understanding seems unneeded in light of the obvious and daily ease with which our effective control of them — primarily through the use of our vision — proceeds with no such thoughts.

But the differences between the attributes of machines and the familiar attributes of humans are never more important to appreciate. In the world of the machine where the coupling of control with the remarkable visual facility of humans is not, outside of human in the loop control, an option, the distinctions between "holonomic" and "nonholonomic" become foundational understandings.

The importance of this discussion is easiest to see by reflecting on the angular-position servomechanism, a paradigm of control nowhere to be found in nature but fundamental to a wide range of engineered systems including those teach-repeat machines that we call robotic arms. Recall from Chapter 2 the "closed loop" of the position servomechanism. Stored into the device's memory, one or a series of targeted angles of rotation are drawn upon for each joint. Feedback, primarily a quantitative measure of the current, actual angle, is compared against this desired or "reference" angle. The actuator or motor responsible for driving a given joint toward its designated, stored angular goals is pushed or restrained according to this angular error using a "control law" or rule that governs the actuator effort in response to discrepancies between reference targets and actual angular measurements.

Within reasonable bounds of variability of load or other kinds of resistance to motion the machines are able to smoothly, quickly and reliably drive toward and, as required, come to rest and sustain, their targeted internal angular positions. The net effect has proven exceedingly useful for "holonomic" mechanisms in a factory context.

To appreciate why the position servomechanism enables holonomic but not nonholonomic robots, consider an example of one of each as shown in *Figure 9-1*. The first is a two-degree-of-freedom, holonomic robotic arm; the second a two-degree-of-freedom robotic wheelchair, a nonholonomic system. Two motors are tasked with actuating both systems; these motors achieve their objective by rotating through to new angular positions. In the case of the arm, much as our muscles would actuate our elbow and shoulder angles, the two motors drive θ_1 and θ_2 of the mechanism. In the case of the wheelchair, something similar occurs but θ_1 and θ_2 are wheel rotations.

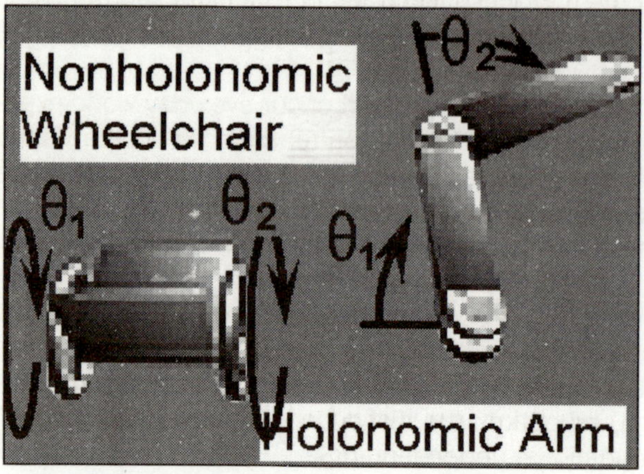

Figure 9-1. Holonomic and Nonholonomic Systems.

Suppose the arm starts in the position shown in the first row of *Figure 9-2* and proceeds to rotate the arm up as shown.

After this motion, an angular-position servomechanism that controls the arm is commanded to return it to the θ_1 and θ_2 positions indicated in the figure, and it succeeds using the unnatural means of angular-position feedback in accomplishing just this, as indicated in the second row of *Figure 9-2*. The plots of angular position histories of *Figure 9-2* are shown in *Figure 9-3*. It is a perfect return to what might have been a human-taught pose of the arm, its original pose.

Consult **http://www.nd.edu/~sskaar/** for an animation of this entire motion, shown in *Figure 9-2* as stills.

Now similar angular histories could also be applied to the wheelchair. And it is certainly true that, if one wanted to (though no one actually does this, for reasons we will see), it would be possible to build an angular-position wheel-control servomechanism θ_1 and θ_2 of the wheelchair. *Figure 9-4* shows in the first row the conversion of a holonomic system to a nonholonomic one. The second and third rows show to the system's movement while rotating θ_1 and θ_2 forward and then backward, respectively.

The point is, return of these angles to an original internal pair of rotations does not guarantee return of the chair to its starting pose. *Figure 9-5* shows the difference between initial and final poses of the nonholonomic wheelchair after applying the same angular history of the robotic arm.

The reason for the difference lies in the mathematical form of the "constraint" between the angles of the wheels and the pose of the mechanisms. The words holo-

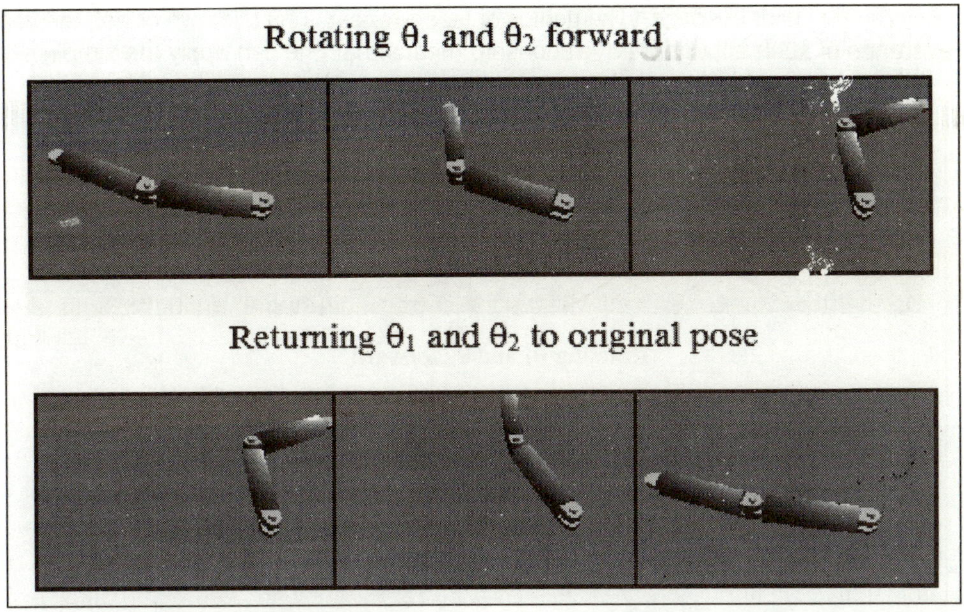

Figure 9-2. Motion of Holonomic System.

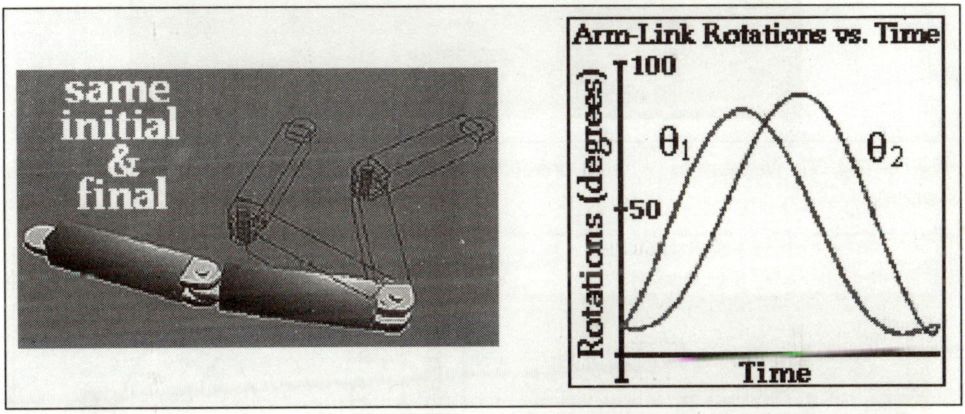

Figure 9-3. Holonomic System and Angle Position Histories of θ_1 and θ_2.

nomic and nonholonomic were applied by pioneers in the field of mechanics to refer to just this "kinematic" distinction. Whereas both systems effect pose changes through internal rotations, for the robotic arm the relationship is algebraic while for the wheelchair it is at most "differential". What that means is that there is a correspondence that can be characterized in advance between θ_1 / θ_2 on the one hand and the location of the positioned tip of the arm on the other. In the language of Chapter 7, it means that there exist (although that chapter claims they are not usually knowable with useful accuracy) relationships g_x and g_y between internal angles and physical position of a point in space. The genius of teach-repeat is that, known or not, the very existence of such an algebraic relationship means that one can apply the simple ser-

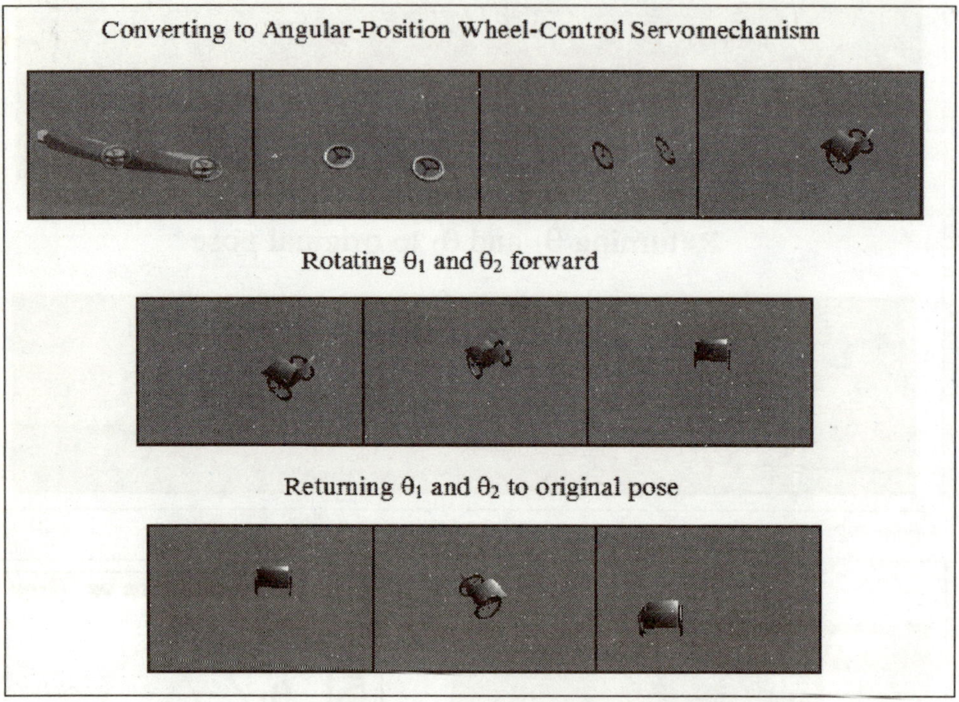

Figure 9-4. Conversion to Wheel-Control Servomechanism and Motion of Nonholonomic System.

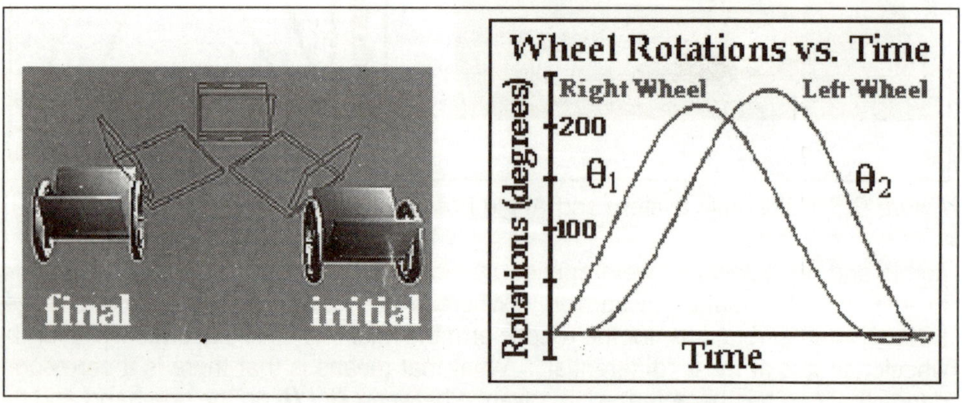

Figure 9-5. Nonholonomic System and Angle Position Histories of θ_1 and θ_2.

vomechanism to cause the arm to return to whatever pose in space a human teacher may have determined got the job done.

But what if you tried this with the wheelchair? As can be seen in *Figures 9-4* and *9-5*, the terminus of the chair, even if you manage to return the wheels to their original internal angles, will not in general be the same as the initial pose. In fact, the net effect of the motion is "path dependent". The relationship, in the end, between terminal joint angles and position or pose within the floor of the chair depends on how you happened to get there – the history of movement that you followed. This nonholonomic characteristic is true of other systems such as yourself and the key-pushing exercise of Chapter 2. It did not bother or upset your ability to accomplish the key-pushing goal because of the way we humans achieve such goals, with ongoing visual feedback. We do not, indeed we can not, use internal-joint feedback in the form of a joint-level servomechanism.

And so it is with the wheelchair. We plan and execute trajectories with no conscious regard paid to the holonomic/nonholonomic distinction. This is what is good about what we can do, but it also relates to what limits us vis-à-vis the machines. The purpose of the present volume is to point the way to artificial systems that have attributes of both the human and the engineered, such as the latter stands today. We want to take further advantage of machines' ability to measure, recall and recover precise internal joint angles without sacrificing the human ability to respond to "as-located". The astute reader may suppose that the difference between the ability of holonomic robots to repeat position with a simple return to internal joint rotations, in contrast with nonholonomic wheeled systems, is due to the likelihood of wheel slip. While it is true that wheel slip can worsen the error in terminal position, this phenomenon is due to the essential characteristics of the nonholonomic constraints. We formalize this discussion below.

In the wheelchair system, understanding the relationship between wheel rotation and the position/orientation of the vehicle requires that we define $\boldsymbol{X}$, $\boldsymbol{Y}$ and $\boldsymbol{\Phi}$. Choosing an arbitrary absolute reference point on the plane of the floor where the wheelchair moves, we define the point $(\boldsymbol{X}, \boldsymbol{Y})$ as the midpoint of the axle of the steering wheels. The angle $\boldsymbol{\Phi}$ determines the orientation of the vehicle with respect to our chosen horizontal axis. These quantities will be more rigorously discussed in Chapter 10. If we denote the five coordinates that determine the state of our system $\boldsymbol{\theta}_1$, $\boldsymbol{\theta}_2$, $\boldsymbol{X}$, $\boldsymbol{Y}$, $\boldsymbol{\Phi}$ by u_1 through u_5, the three nonholonomic constraints have the general form:

$$\sum_{k=1}^{5} a_{jk}\, du_k + a_{j0}\, dt = 0, \quad j = 1, 2, 3 \tag{9.1}$$

where the coefficients a_{jk} may, in general, be functions of u_1 through u_5, as well as time. In our example system, and in the class of systems considered in this book, time does not appear explicitly in these coefficients; moreover, $a_{j0} = 0, j = 1, 2, 3$.

With a holonomic robot, constraints among the "internal" coordinates (joint rotations θ_1 and θ_2) and "external" coordinates $\mathbf{X}$, $\mathbf{Y}$, $\boldsymbol{\Phi}$ can also be written in the form,

(9.2) $$\sum_{k=1}^{5} a_{jk}\, du_k = 0, \quad j = 1, 2, 3$$

where $\mathbf{X}$, $\mathbf{Y}$, $\boldsymbol{\Phi}$ represent the position and orientation of the tip of the end member. Note that the robotic arm shown in *Figure 9-1* can only move in two dimensions.

However, only in the holonomic case these three constraints are integrable to the algebraic form:

(9.3) $$f_1\,(u_1, u_2, u_3, u_4, u_5) = 0, \quad j = 1, 2, 3$$

There are two implications of the above contrast between holonomic and nonholonomic systems:

1) Due to the algebraic nature of constraints of equations (9.3), two inputs (θ_1, θ_2) are only able to bring independently specifiable terminal positions of two of the three quantities of interest ($\mathbf{X}$, $\mathbf{Y}$, $\boldsymbol{\Phi}$). For the nonholonomic case, all three can be controlled via the two inputs θ_1 and θ_2. Thus, for a given number of controlled outputs, fewer degrees of freedom may be required for the nonholonomic system.

2) The outcome of a maneuver ($\mathbf{X}$, $\mathbf{Y}$, $\boldsymbol{\Phi}$) depends upon the terminal value of inputs θ_1 and θ_2 only. By contrast, the outcome for the nonholonomic system is path-dependant. If an error in tracking the reference rotations occurs, return to the reference (θ_1, θ_2) path will cause a holonomic robot to terminate at the desired pose, whereas, in general, a nonholonomic robot will terminate at an incorrect pose, under the same circumstance.

The characteristic of returning a robot's internal joint rotation to a predetermined configuration is the requisite for *repeatability*. Repeatability is the basis of teach-repeat, the main means of control of industrial robots. Teach-repeat is a very simple technique, but it is its very simplicity and reliability that makes it so appealing and widely used. It takes advantage of the amazing human intuition to construct a prototype trajectory, and fully exploits the ability of a robot to reproduce it automatically. Paradoxically, teach-repeat can be accomplished with little or no reference to many of the most prominent sciences studied in robotics, such as kinematics or kinetics. Joint level control can be achieved while being more or less ignorant of system dynamics due to excellent design control algorithms and precise devices like rotational encoders.

Since repeatability depends on the holonomic nature of the system, it would seem that teach-repeat can only be used on systems of this type. However, teach-repeat can be extended to nonholonomic robots. As always, there is tradeoff: to take advantage to this technique we lose some of the simplicity of the approach since we need the aforementioned sciences as well as estimation methods to apply it. However, we gain the property of repeatability on wheeled robots.

This topic is the subject of our next chapter, *Extending Teach-repeat to Nonholonomic Robots.*

CHAPTER 10

EXTENDING TEACH-REPEAT TO NONHOLONOMIC ROBOTS

"When the impossible has been eliminated, whatever remains, however unlikely, must be the truth." Such was the wisdom of Sir Arthur Conan Doyle's Sherlock Holmes. For the present technological objective - usefully versatile navigational autonomy for indoor, wheeled robots - this maxim might be restated: "When the alternatives have been tried and found wanting, the technology left standing - estimation-based extension of teach-repeat to nonholonomic robots - may prove a remarkably competent basis for achieving most real-world ends."

In Chapter 2, we discussed how the teach-repeat technique has been employed with great success in holonomic systems. This paradigm exploits the fact that a human teacher has great intuition on planning a potential trajectory, and that a holonomic system (for example, a robotic arm) can automatically replicate said trajectory thousands of times. If higher precision for a certain task is needed, the human operator can simply teach more intermediate points in that particular region of the path. In short, teach-repeat relies on the capacity to return the robot's internal "joint configuration" to the one taught or pre-planned. Due to its simple but reliable nature, the teach-repeat approach is widely employed in industry, particularly in assembly lanes.

In Chapter 9, we discussed the "kinematic" distinction between holonomic and nonholonomic systems. In a holonomic system, returning to the original state means returning to the initial pose. In contrast, motion in a nonholonomic system is path dependant, which means that returning to the original joint configuration does not result in returning to the original pose. This chapter is concerned with extending the teach-repeat paradigm to nonholonomic systems. This extension is not trivial, since it requires estimation methods unlike its holonomic counterpart. This section of the book describes in detail the step by step implementation of an application that uses the teach-repeat paradigm for navigation.

The project used as an example, a fully autonomous wheelchair prototype, was developed at the Dexterity, Vision and Control Laboratory at the University of Notre Dame. This is a project in collaboration with the E. J. Hines Jr. VA Hospital in Chicago, where most of the testing with subjects was conducted. The current name of the prototype is *Computer Controlled Power Wheelchair Navigation System*, or *CPWNS*. *Figure 10-1* shows a picture of the *CPWNS* vehicle.

The goal of this project is to give a certain level of autonomy to people who have lost the use of their legs and lack the necessary coordination to manipulate the joystick or other user-input device of a motorized wheelchair. This situation can happen, for example, with a crippling spinal cord injury. A recent case that received attention in the media was the death of actor and activist Christopher Reeve, who broke his neck while riding a horse in 1995. While alive, he used his celebrity status to bring

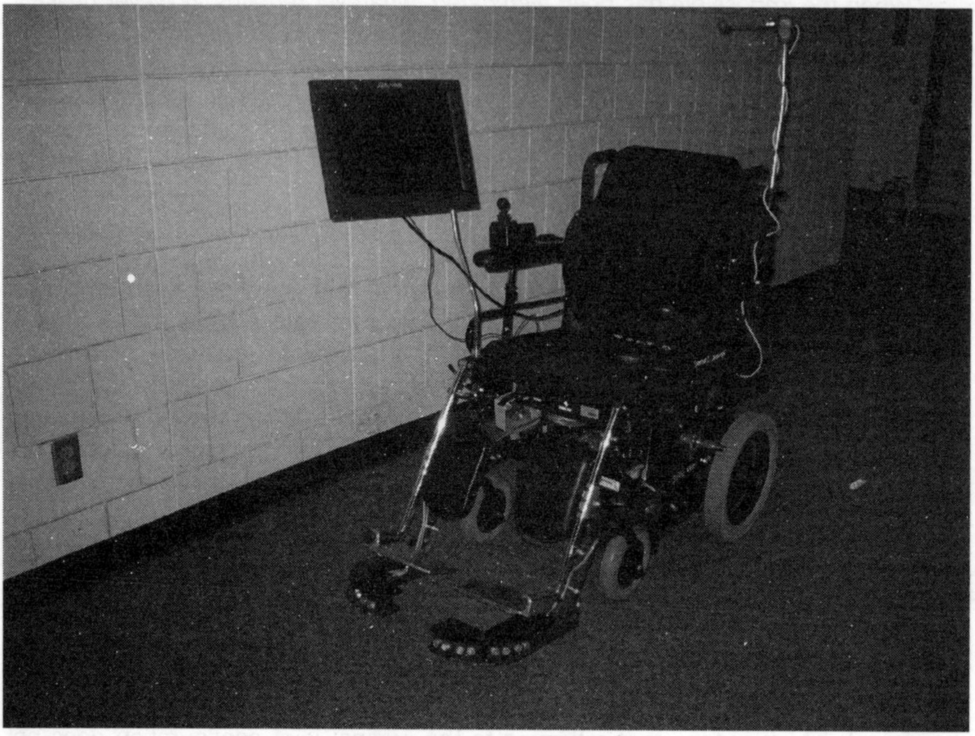

Figure 10-1. The *CPWNS*.

attention and a recognizable face to the plight of people who are paralyzed. They are, in fact, prisoners in their own bodies.

A severe state of disability can also be the result of malady, like multiple sclerosis or motor neuron disease. Stephen Hawking, to mention another famous case, has Amyotrophic Lateral Sclerosis. This disease is characterized by a progressive degeneration of the motor cells in the spinal cord and the brain. The symptoms include paralysis, impaired hearing and difficulty to swallow and breathe. However, in most cases, it does not affect memory or mental faculties. Hawking's condition has been captured in the excellent documentary *A Brief History of Time* (1991) directed by Errol Morris.

Finally, people can develop retinopathy and/or neuropathy due to sequelae associated with diabetes. Retinopathy, a kind of eye disease, is a serious complication that may lead to diminished vision, contraction of the visual field, or blindness. Other complications of diabetes, cataracts and glaucoma, can also lead to total vision loss. Neuropathies or nerve diseases are among the most frequent complications of long-term diabetes. They commonly affect the legs and feet. Patients in an advanced stage of diabetes can be both blind and with severe limitation of the use of their legs. At the Hines VA Hospital we met such a patient, a blind woman who steered her motorized wheelchair with one hand and waived a cane with the other. She used this cane as a makeshift "bumper sensor" to find her way around a room, and could actually move around reasonably well.

In general, patients with such a severe disability find it difficult to steer a motorized wheelchair in a confined environment. In a house where maneuvering space is limited, somebody with diminished visual skills or unsteady hand coordination could face a formidable challenge. Especially problematic are the actions of approaching furniture without collisions and crossing the threshold of a doorway.

These complex actions require a certain level of hand-eye coordination that individuals with high-level spinal cord injury, multiple sclerosis, or brain injury do not have. For some of these patients, learning how to use a motorized wheelchair can take months or years.

A clinical survey conducted by the Rehabilitation Research & Development and Spinal Cord Injury at the Hines VA Hospital concluded that 18%- 26% of their patients confined to a manual wheelchair could not operate a motorized one. The study concluded that no independent mobility options for these patients existed at the time. This is a population that could benefit from an autonomous wheelchair that can receive simple commands and navigate unaided in a complex environment.

Autonomous wheelchairs are a subset of autonomous mobile robots that must have the following characteristics: maneuverability, navigation, control, safety; these set them apart from other more general autonomous mobile robots. The issue of safety is particularly important, since it would not be acceptable for the vehicle to ram or graze walls or furniture while a paralyzed subject is riding it.

If, as an exercise, you do a Google search for 'Autonomous wheelchairs', thousands of hits are returned. Among them, *NavChair* developed at the University of Michigan, *Rolland* developed at the University of Bremen, the *TechWeb* article 'Robots on Parade', etc. However, most of the referenced systems are only semi-autonomous, at least in terms of navigation, or they are something else entirely.

The 'something else entirely' category belongs to what is called "Smart wheelchairs". These particular systems are designed with more concern about the human-machine interaction: sensors that can receive input from the user via eyebrow or eyelid movements, bite sensing, voice activation, etc. The emphasis is in creating an intuitive interface for a disabled user to communicate commands to an onboard computer. However, these vehicles still need user interaction for navigation. Some of these systems are the *RobChair* developed at the University of Coimbra in Portugal and the

SIAMO system developed at the University of Alcalá in Spain. A famous commercially available example is the *iBOT*, a wheelchair that climbs stairs and that got FDA approval in August 2003. It was developed by Dean Kamen, who also invented the Segway scooter. In the case of the *iBOT*, the main feature is the stair-climbing capacity of the system, not autonomous navigation.

Other robotic wheelchairs have semi-autonomous navigation. That is, they are able to perform certain navigational tasks without user input: following a wall, avoiding an obstacle, moving in straight line, following a corridor. For more complicated maneuvers they need user input. This input can be given in terms of a shaky joystick movement, which the control can then filter and 'interpret' into the desired direction to travel. These systems can supplement this motion with obstacle detection/ avoidance, which is useful when trying to cross the threshold of a door. In some cases, the working paradigm is behavior-based, which means that the algorithm can receive user input AND follow a wall AND avoid obstacles, etc. Examples of this kind of system are *Rolland* – developed at the University of Bremen, Germany, *VAHM* – developed at the University of Metz, France and *NavChair* - developed at the University of Michigan.

The problem with the semi-autonomous approach is that it does not solve the main issue of navigation. These are partial solutions that work in particular situations. For example, following a wall or a corridor might be useful in a hospital or a large office building, but not viable in a cramped room with a collection of small and large bodies.

The navigational problem is extremely difficult to solve in general, cluttered, realistic environments. It is so difficult, in part, because it relies on the solution of other general issues that have not been solved in a satisfactory or general form. Some applications, like the vacuum cleaner robot *Roomba*, bypass these unresolved issues by using random walks for navigation. This is also the paradigm used in the autonomous lawnmower robots. In those cases, the purpose of navigation is to completely cover an area, either to mow it or to clean it. However, this is not an acceptable solution in the case of autonomous navigation, since the wheelchair should not move at random in a room until it arrives (eventually, one would hope!) to the desired destination.

Calculating a trajectory connecting two points can be quite a difficult task. There are several parameters and physical limitations of what constitutes a "good" trajectory. This is what is known in robotics literature as the *path planning* problem. To solve it, some mobile robotic systems need to have detailed maps of the environment, and devote a large part of their processing capability to create and update these maps. One such approach, Simultaneous Localization and Mapping (SLAM), has become quite popular lately in the robotics community. Other approaches simply ignore trajectory planning and substitute it with some behavior-based scheme. An example of this type of solution is the previously mentioned *Follow Wall* algorithm. As it name implies, the system simply follows a wall until a certain condition is reached. The problems and limitations of behavior-based approaches have been discussed in Chapter 4.

A possible, simplistic solution to the path planning problem would be to put a magnetic strip on the floor of an environment, and have the wheelchair follow it. No matter how cluttered or complex the workspace, the wheelchair could arguably take its user to his destination as long as the strip can be detected. The system could be made robust enough, so that following the magnetic strip would be akin to have the wheelchair mounted on a rail.

However, this idea poses several practical problems. First, the movement would be limited to the marked paths. Second, these paths cannot physically overlap without creating confusion in the system, or creating the need for additional information.

Now, imagine that we could use a virtual magnetic strip. It would still work as some sort of rail that the vehicle could track, but the strip itself would not need to be installed on the floor. The trajectory could somehow be programmed in the system. Since the strip would be virtual, a multitude of paths could be included, even ones that overlap in one or several points. The goal is to make these virtual paths as reliable as the physical ones. *This is one of the main ideas behind extending teach-repeat to nonholonomic systems.* The teach-repeat idea bypasses the hard problem of trajectory planning. It avoids the costly calculations of "classic" approaches, but does not ignore the problem, unlike behavior-based schemes. Rather, it puts the burden of planning on a human teacher, as in the holonomic case.

A direct consequence of extending teach-repeat to nonholonomic systems is *to fully exploit the remarkable human capacity to create trajectories. Figures 10-2* and *10-3* illustrate the teach-repeat steps. Later in the chapter, we will see that there is an extra step between the first and second episode. This step is related to streamlining the points generated by the teaching episode.

Figure 10-2. Teaching the trajectory.

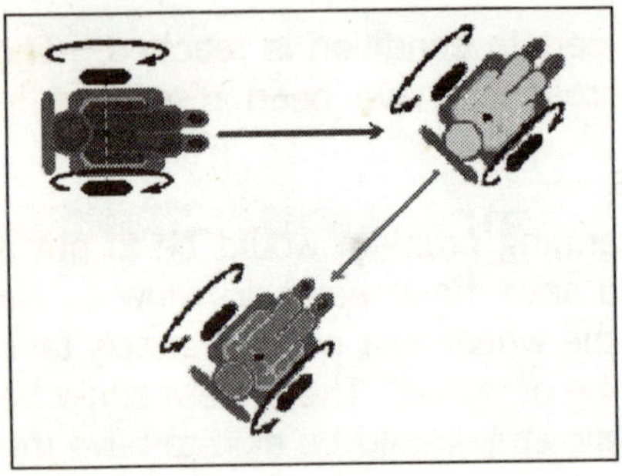

Figure 10-3. Tracking the trajectory.

Before proceeding any further, let us recapitulate: the wheelchair is walked through the desired path by a human teacher, while gathering pose information. These collected readings will allow it to repeat the trajectory. When repeating the maneuver, the vehicle should be as safe as if it were running along a rail. Ideally the human teacher will need little training and no engineering experience to be able to set up and operate the system. We will also see that as a byproduct of using the teach-repeat approach, the system does not need a complicated representation of the environment.

Now, let us take a look at the physical side of the *CPWNS*:

The prototype consists of a standard power wheelchair, an onboard PC with an Intel Pentium 4 Processor at 1.80 GHz, powered by a 14.1-Volt lead acid gel battery. To measure the wheel rotation, the vehicle has two rotary incremental encoders connected to a data acquisition board. The user interacts with the system using a biting switch, a chin switch, or voice activation. All the different switches are connected through the parallel printer port of the desktop. A monitor displays the available commands, and voice encoding enables the wheelchair to "talk" to the patient. The system has also cameras and sonar transducers. We will describe that part of the system when we introduce the concepts of estimation and obstacle detection. *Figure 10-4* shows a user riding the *CPWNS* at the Hines, VA hospital.

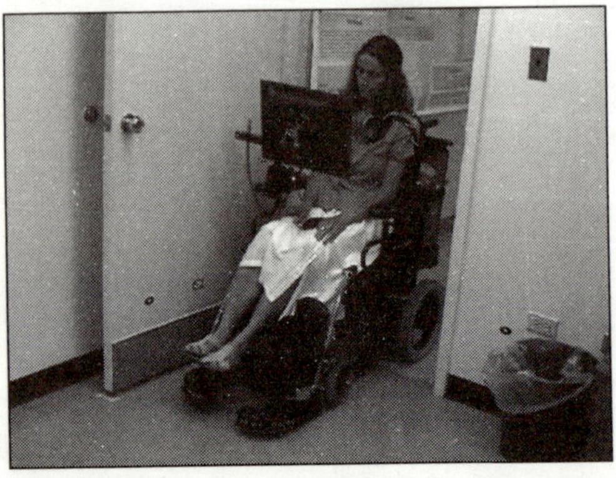

Figure 10-4. User riding the *CPWNS*.

Observe the wheelchair schematic shown in *Figure 10-5.* We define the variables θ_1 and θ_2 as the left and right rotations of the steering wheels, respectively. We let X, Y be the Cartesian coordinates of ***A*** – the midpoint of the axle. The angle ϕ measures the orientation of the system with respect to the X axis of the fixed floor coordinates. The coordinates X, Y are absolute with respect to a predetermined reference point.

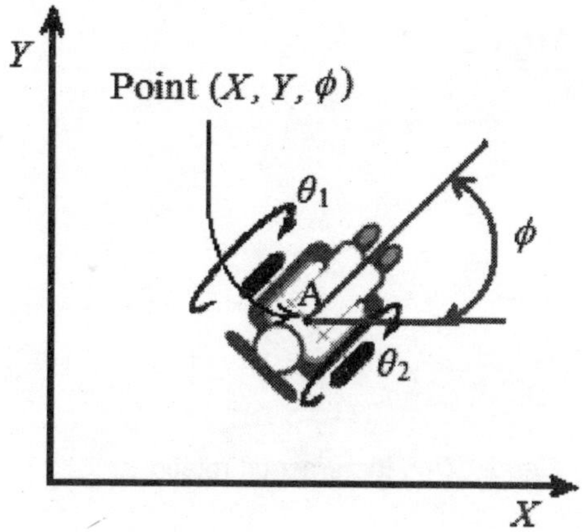

Figure 10-5. System Definition.

In Chapter 9 we saw that the two quantities (θ_1, θ_2) completely determine the other quantities of interest: (X, Y, ϕ). In the *CPWNS*, θ_1 and θ_2 are measured using the rotary encoders.

Consider now the motion of the wheelchair towards a table, as indicated in *Figure 10-6.* During the teaching episode, the chair undergoes motion such that the mid-point ***A*** between the two wheels moves from the starting point ***C*** to the destination point ***D***. During the teaching episode the chair is guided by a human operator. It can be either pushed or controlled by joystick.

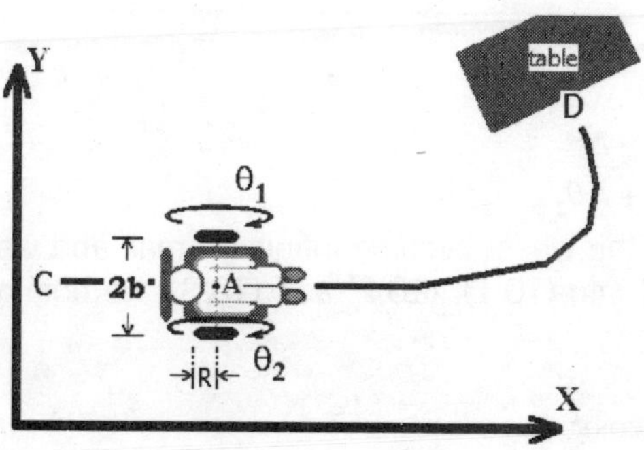

Figure 10-6. Taught Path from Position ***C*** to Destination ***D***.

The two parameters of the system are the radius R of each drive wheel, and the distance b from the wheel to the midpoint of the axle. Assuming that the wheels do not slip, consider a very small increment in the two wheel rotations, $\Delta\theta_1$ and $\Delta\theta_2$. This change results in a small increment in the position of the two ends of the axle of $R\Delta\theta_1$ and $R\Delta\theta_2$. *Figure 10-7* shows the linear distances traveled by both drive wheels and the midpoint of the axle.

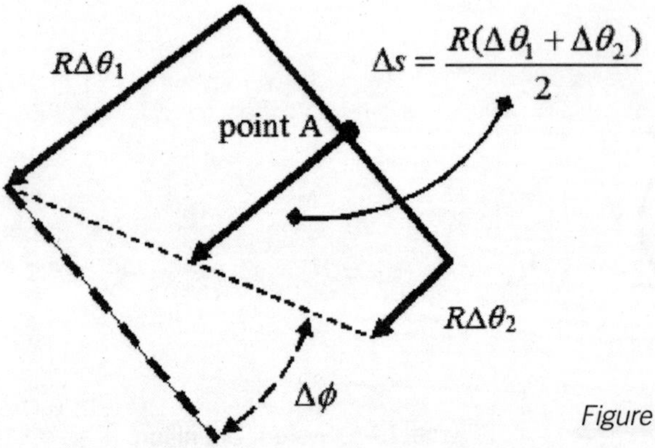

Figure 10-7. Incremental rotations.

Calling s the distance traveled, we can write the approximate equations of the change in position and orientation of point ***A***, the midpoint of the axle.

(10.1) $$\Delta\phi \approx \frac{R(\Delta\theta_1 - \Delta\theta_2)}{2b}$$

(10.2) $$\Delta X \approx \Delta s \cos\phi$$

(10.3) $$\Delta Y \approx \Delta s \sin\phi$$

We define the quantities u and α as follows:

(10.4) $$\alpha = \frac{s}{R} = \frac{\theta_1 + \theta_2}{2}$$

(10.5) $$u \approx \frac{\Delta\theta_1 - \Delta\theta_2}{\Delta\theta_1 + \Delta\theta_2}$$

If we take the limit as the rotations of the wheel become infinitely small and we substitute the equations (10.4) and (10.5) into (10.1), (10.2) and (10.3), we obtain the kinematic equations of motion.

(10.6) $$\frac{dX}{d\alpha} = R\cos\phi$$

(10.7) $$\frac{dY}{d\alpha} = R\sin\phi$$

(10.8) $$\frac{dY}{d\alpha} = R\sin\phi$$

Note that in this new form u and α are the independent variables of the system, instead of θ_1 and θ_2. The variable u gives a relation of the differential rotations of the two drive wheels, and is in fact the control steering of the vehicle while tracking. The independent variable α is defined in terms of the individual wheel rotations, instead of the more familiar variable time. The reason why this is convenient will be clear when we introduce the need for estimation.

We can group the equations of motion (10.6) - (10.8) in vector form

(10.9) $$\frac{d\underline{x}}{d\alpha} = \underline{f}(\underline{x},u)$$

where

(10.10) $$\underline{x} = [X,Y,\phi]^T$$

The vector equation (10.9) can be integrated forward from the starting position ***C*** of the wheelchair (X_C, Y_C, ϕ_C) to produce ongoing values of u and α. The numerical integration that produces the values of the position and orientation of the system from wheel-encoder readings is called odometry or *dead-reckoning*.

Note that in order to derive the equations of motion and to define u and α, we need to choose a direction of movement of the wheels. For the particular case of equations (10.4)-(10.8), the vehicle was assumed to be moving forward. This means that both wheels are moving in the same direction, though not necessarily at the same speed. If the chair is moving backwards, the definitions of u and α in equations (10.4) and (10.5) change in sign. The derivation of the equations of motion is analogous. One only has to be careful in the propagation of the sign in further calculations.

So far we know how to model the movement of the vehicle when both wheels move in the same direction, backwards or forwards. At this stage the wheelchair could move like a car, with the limitations that a car-like movement entices. However, a real wheelchair can move its steering wheels in opposite directions. That is, unlike a car, it can do a purely rotational movement, or a pure *pivot*. However, as the vehicle approaches a movement that is pure rotation, the control variable u tends to grow, as the denominator tends towards zero. In pure rotational motion $\Delta\theta_1 = -\Delta\theta_2$, and the vehicle just spins into place with no translation. Looking at equation (10.5), one sees that as $\Delta\theta_1 \rightarrow -\Delta\theta_2$, $u \rightarrow \infty$. This is not a useful description of the state of the system. We mentioned earlier that u is related to the control of the system, which means that the wheelchair needs an infinite steering signal to be able to execute a pure pivot. Moreover, notice that the independent variable α approaches a constant in equation (10.4).

Pure rotational-motion capacity is desirable in a vehicle of this kind for a variety of reasons:

- A chair that can only move like a car cannot do sharp turns, and has to compensate with several linear movements what could be done with one single rotation.
- Realistic working areas are normally very crowded, and space is at a premium. They could include small corridors, narrow doorways, bedrooms with furniture, bathrooms and kitchens. Some of these environments are impossible to navigate without pivoting.
- The limitation given by the impossibility of executing a purely rotational movement limits the range of trajectories that a human could teach to the vehicle.

Figure 10-8 is the analogue of *Figure 10-7* in a situation when the wheels are moving in opposite directions. The wheelchair is rotating counterclockwise.

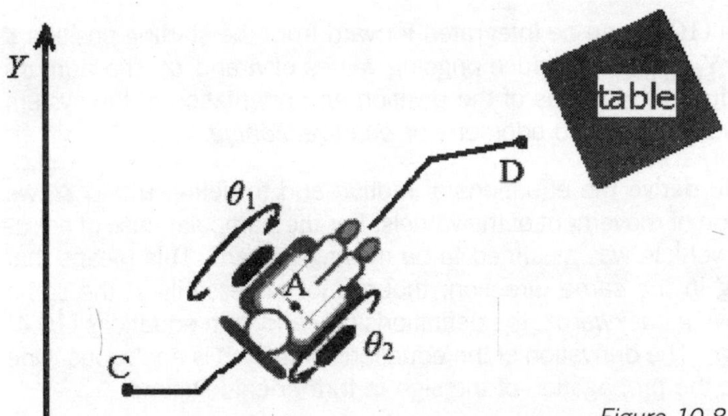

Figure 10-8. Taught path with wheels moving in opposite direction.

For the reasons that we mentioned, new definitions for α and u are needed for the case $\Delta\theta_1 > 0$ and $\Delta\theta_2 < 0$:

(10.11)
$$\alpha^* = \frac{s^*}{R} = \frac{\theta_1 - \theta_2}{2}$$

(10.12)
$$u^* \approx \frac{\Delta\theta_2 + \Delta\theta_1}{\Delta\theta_1 - \Delta\theta_2}$$

This behavior is called *pivot right*. Note that $u^* = 1/u$. Making this change of variables yields a new set of kinematic equations of motion

(10.13) $$\frac{dX}{d\alpha^*} = R\,u\cos\phi$$

(10.14) $$\frac{dY}{d\alpha^*} = R\,u\sin\phi$$

(10.15) $$\frac{d\phi}{d\alpha^*} = -\frac{R}{b}$$

Since there is no indication that the wheels are moving at the same speed, the vehicle is not necessarily doing pure rotational motion. In fact, it can reach its destination ***D*** from the point ***C***. It is useful to think of the variable s^* as an *angular* distance traveled, as opposed to the *linear* distance traveled defined in equation (10.4). The term linear can be misleading in this context, since it might suggest that the vehicle is moving in a straight line, which is not necessarily the case.

The last type of motion corresponds to $\Delta\theta_1 < 0$ and $\Delta\theta_2 > 0$. In this case, the vehicle turns in a clockwise way. This rotational behavior is called *pivot left*. The definitions of u^* and α^* change slightly,

(10.16) $$\alpha^* = \frac{\theta_2 - \theta_1}{2}$$

(10.17) $$u^* \approx \frac{\Delta\theta_2 + \Delta\theta_1}{\Delta\theta_2 - \Delta\theta_1}$$

The equations of motion have the same form as in (10.13) – (10.15), with some sign differences. This is analogous to the forward and backward motions discussed earlier. *Figure 10-9* shows the four different kinds of movement of the wheelchair, along with the different definitions of the distance traveled α and of the parameter u.

In the figure, the sub indices ***F***, ***B***, ***R***, ***L*** indicate forward, backward and pivots right and left, respectively. They have been added for clarity. We saw that $\alpha_B = -\alpha_F$ and $u_B = -u_F$. In rotational movement $\alpha^*_L = -\alpha^*_R$ and $u^*_L = -u^*_R$. The relationship between the definitions of parameter u is: $u_F = 1/u^*_R$ and $u_B = 1/u^*_L$. We can simply express them by

(10.18) $$|u| = \frac{1}{|u^*|}$$

Since we apply an absolute value to the equation, the sub indices can be dropped. Equation (10.18) essentially shows the type of movement of the vehicle while it is being taught a trajectory. The more rectilinear the trajectory, the more u tends to zero and u^* to infinity. Conversely, the more rotational the trajectory, the more u^* tends to zero and u to infinity. Since both definitions of u are related to the control,

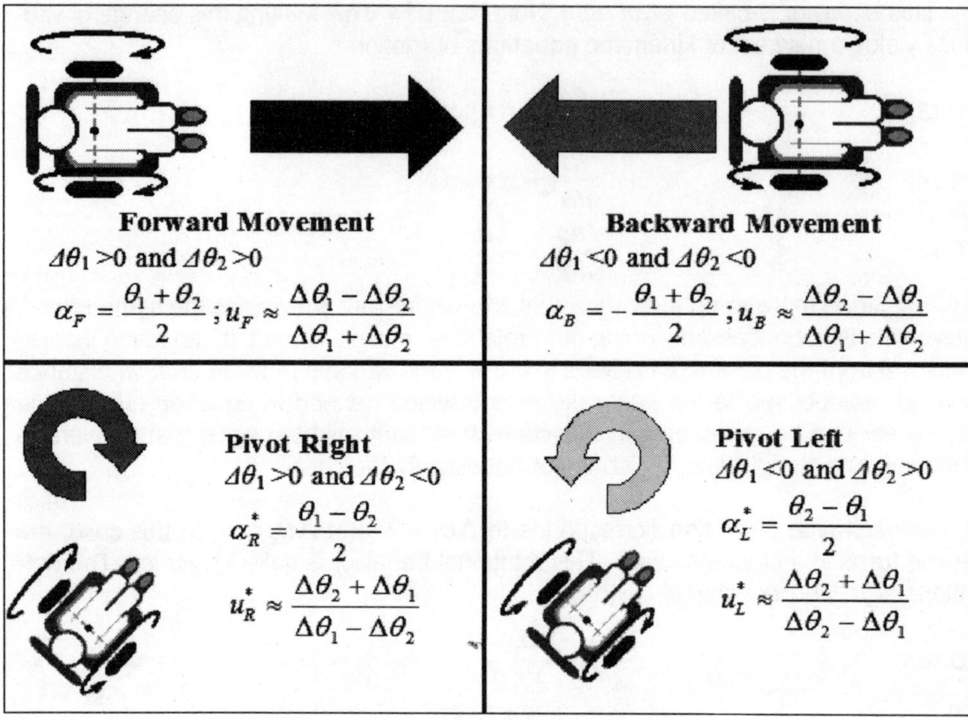

Figure 10-9. Four modes of movement.

this parameter is extremely useful to describe the movement. In other words, it helps us choose which definitions of $\frac{d\underline{x}}{d\alpha} = \underline{f}(\underline{x},u)$ and $\underline{x} = [X, Y, \phi]^T$ we need to integrate using the encoder readings.

In principle, we could take the wheelchair and 'walk it' through a series of poses. The position of the system could be calculated by knowing the initial position with respect to a fixed reference, and integrating the right set of definitions of $\frac{d\underline{x}}{d\alpha}$. These poses could be stored in some useful form, we still have not seen how, and then reproduced during the tracking step. This is the way a holonomic system works. However, in a nonholonomic system we cannot simply apply forward integration. The main reasons for this limitation are:

- In order for a nonholonomic system to be taught a path effectively, it must be able to determine its current pose. This is known as the *location problem.* This problem is difficult to solve because of the differential nature of its input-output equations.

- Dead-reckoning is very inaccurate over long paths due to small, unavoidable errors in initial conditions, the system model, wheel slippage, and numerical integration. These errors grow in magnitude as the integration progresses.

A solution is to provide observations to obtain or correct the pose information. Among several possibilities that could have been incorporated in our system, we chose a well-known technique called *estimation*.

Of course, this is not the only possible approach. For example, one could eliminate the *dead-reckoning* part of the algorithm, and use only observations to triangulate the position of the vehicle. Such systems exist. Their drawback is that they must have a certain minimum of reference points or beacons, and that they are only as good as their last measurement. Another possibility is to improve as much as possible the odometry integration. The parameters of the system could be measured as carefully as possible, the model could take into account heat dilation of the components, wheel slippage, etc. This approach would be akin to doing a very careful calibration of the system. Such systems normally require extensive setup, need complicated models, and are very vulnerable to an unforeseen change in conditions – for example a different floor texture. Other paradigms like SLAM or behavior-based robotics have been discussed at length in earlier chapters.

The main advantage of estimation is that it takes into account both the information of the model *(dead-reckoning)* and the one provided by external observations. A weight can be assigned to new information, depending on the accumulated history of the pose measurements. In some cases, a new information can even be rejected on that basis. In the case of nonholonomic systems, estimation is typically used to combine the odometry calculation with observations related to the current pose of the system. The observations can be obtained though different types of devices: sonar, vision, laser range-sensing, etc. The Extended Kalman Filter is a well known algorithm widely used to integrate sensing information.

We will now see how the Kalman Filter can be applied to our wheelchair system. The discussion that follows has more in common with a cookbook recipe, and is not a formal justification.

Let us start with a slightly different version of the state equations (10.9) and (10.10).

(10.19)
$$\underline{x}(\alpha) = \left[X(\alpha), Y(\alpha), \phi(\alpha)\right]^T$$
$$\frac{d\underline{x}(\alpha)}{d\alpha} = \underline{f}(\underline{x}(\alpha), u(\alpha)) + \underline{w}(\alpha)$$

A process-noise vector $\underline{w}$ has been added to $\underline{f}$ to account for errors in the nominal nonholonomic kinematics. The process $\underline{w}$ is assumed to be zero-mean, with Gaussian distribution uncorrelated in α, and random. It has an associated covariance matrix $\mathbf{Q}$. The proposed set of equations (10.19) is not essentially correct, since it assumes that the error in the kinematics equations can be represented by white noise. However, the white noise assumption can be used effectively as a representation of the errors in the system model and in the measurements of the physical system. The α parameter has been added in (10.19) to stress the fact that the equations are only dependant on this variable and on u. The definitions of α and u depend on the mode of movement. That is, in the following discussion α stands for either α_F, α_B, α^*_L, or α^*_R. Similarly, u means u_F, u_B, u^*_L, or u^*_R. The same criterion applies for the vector equations $\underline{f}$ and $\underline{x}$.

Assume that the observations of the system can be modeled by the following vector equation,

(10.20)
$$\underline{z}(\alpha_a) = \underline{h}(\underline{x}(\alpha_a)) + \underline{v}(\alpha_a) .$$

Note that the vector equation $\underline{z}(\alpha_a)$ is only valid at discrete values of the independent variable α_a, when an observation is obtained. The model of the sensing device is specified by $\underline{h}$ called the *observation equation*, with its associated a measurement noise vector $\underline{v}$. The process $\underline{v}$ is also assumed to be zero-mean, random, with Gaussian distribution. It has an associated covariance matrix $\mathbf{R}$. The process noise $\underline{w}$ and the measurement noise $\underline{v}$ are assumed to be uncorrelated. It would be difficult to argue that the assumption of uncorrelated white noise for $\underline{w}$ and $\underline{v}$ is valid. However, the resulting algorithm acknowledges that there are errors both in the system model and in the measurements.

The estimate of the state is represented by $\hat{\underline{x}}(\alpha)$. By this we mean that we do not know the "true" state $\underline{x}(\alpha)$. The estimation error covariance matrix of the state is $\mathbf{P}$, and it is given by

(10.21)
$$\mathbf{P}(\alpha) = E[(\underline{x}(\alpha) - \hat{\underline{x}}(\alpha))\,(\underline{x}(\alpha) - \hat{\underline{x}}(\alpha))^T] .$$

Here E is the expected value of the process. The diagonal terms of the matrix $\mathbf{P}$ represent the variances (that is, the squares of the standard deviations) of the estimation errors of the state. For example the first diagonal element of $\mathbf{P}$, P_{11}, represents the variance of the error of the position estimate of the vehicle in the X direction. For nonlinear systems, $\mathbf{P}$ is only an approximate mean square error and not a true covariance. In both cases it gives a good indication of the general trend of the estimation-error variances of the state.

The Extended Kalman Filter algorithm works in the following way:

1) The state estimates $\hat{\underline{x}}(\alpha) = (\hat{X}(\alpha), \hat{Y}(\alpha), \hat{\phi}(\alpha))^T$ and the estimate covariance matrix $\mathbf{P}(\alpha)$ are initialized at $\alpha = 0$. This means that the initial estimate of the pose $\hat{\underline{x}}(\alpha)$ has to be known, along with the estimates of the variances of the errors for each coordinate. The noise covariance matrices $\mathbf{Q}$ and $\mathbf{R}$ also need to be initialized.

For example, in the *CPWNS* project we use $P_{11}(\alpha_0) = \mathrm{Var}(\hat{X}_0) = 1.5\ \mathrm{in}^2$, $P_{22}(\alpha_0) = \mathrm{Var}(\hat{Y}_0) = 1.5\ \mathrm{in}^2$, $P_{33}(\alpha_0) = \mathrm{Var}(\hat{\phi}_0) = 0.1\ \mathrm{rad}^2$, with all the other values of $\mathbf{P}$ equal to zero. In addition, we use a generic $Q_{11}(\alpha_0) = \mathrm{Var}(\mathrm{Noise}\ X_0) = 2.5\ \mathrm{in}^2$, $Q_{22}(\alpha_0) = \mathrm{Var}(\mathrm{Noise}\ Y_0) = 2.5\ \mathrm{in}^2$, $Q_{33}(\alpha_0) = \mathrm{Var}(\mathrm{Noise}\ \phi_0) = 0.01\ \mathrm{rad}^2$ with all the values of $\mathbf{Q}$ equal to zero. These values need only be feasible, and indicate how much error we think we have in the initial state. Using observations, the Kalman Filter will correct our initial assumptions.

2) When observations are not available, the state estimates and the covariance matrix are propagated by numerically integrating the differential equations

$$\begin{aligned} \frac{d\hat{\underline{x}}(\alpha)}{d\alpha} &= \underline{f}(\hat{\underline{x}}(\alpha), u(\alpha)) \\ \frac{d\mathbf{P}(\alpha)}{d\alpha} &= \mathbf{F}(\hat{\underline{x}}(\alpha), u(\alpha))\mathbf{P}(\alpha) + \mathbf{P}(\alpha)\mathbf{F}^T(\hat{\underline{x}}(\alpha), u(\alpha)) + \mathbf{Q}(\alpha) \\ \mathbf{F}(\hat{\underline{x}}(\alpha), u(\alpha)) &= \left.\frac{\partial \underline{f}(\underline{x}, u)}{\partial \underline{x}}\right|_{\substack{\underline{x}=\hat{\underline{x}}(\alpha) \\ u=u(\alpha)}} \end{aligned} \tag{10.22}$$

This procedure is the *dead-reckoning* part of the algorithm. Note that since is positive definite, it grows at every integration step. The variance of our error grows larger and larger as the position and orientation of the vehicle becomes more uncertain, since the system moves with no observations to provide it with feedback. Note that this intuitive result is a consequence of using α (the distance traveled) rather than time as the independent variable for purposes of estimation. If time was the independent variable and the human teacher paused during teaching of a path, variances would continue to grow between observations despite the fact that there is no motion of the chair.

3) The state estimates and the estimation error covariance matrix are updated at α_a when a new observation is acquired:

$$\underline{\hat{x}}(\alpha_a \mid \alpha_a) = \underline{\hat{x}}(\alpha_a) + \mathbf{K}(\alpha_a)[\underline{z}(\alpha_a) - \underline{h}(\underline{\hat{x}}(\alpha_a))]$$

(10.23) $$\mathbf{P}(\alpha_a \mid \alpha_a) = [\mathbf{I} - \mathbf{K}(\alpha_a)\mathbf{H}(\underline{\hat{x}}(\alpha_a))]\,\mathbf{P}(\alpha_a)$$

$$\mathbf{K}(\alpha_a) = \mathbf{P}(\alpha_a)\mathbf{H}^T(\underline{\hat{x}}(\alpha_a))[\mathbf{H}(\underline{\hat{x}}(\alpha_a))\mathbf{P}(\alpha_a)\mathbf{H}^T(\underline{\hat{x}}(\alpha_a)) + \mathbf{R}(\alpha_a)]^{-1}$$

$$\mathbf{H}(\underline{\hat{x}}(\alpha_a)) = \left.\frac{\partial \underline{h}(\underline{\hat{x}}(\alpha_a))}{\partial \underline{x}}\right|_{\underline{x}=\underline{\hat{x}}(\alpha_a)}$$

The quantities $\mathbf{P}(\alpha_a|\alpha_a)$ and $\underline{\hat{x}}(\alpha_a|\alpha_a)$ are the updated estimate vector and covariance matrix, respectively. These are called *a posteriori* values of $\mathbf{P}$ and $\underline{\hat{x}}$. We denote by $\mathbf{P}(\alpha_a)$ and $\underline{\hat{x}}(\alpha_a)$ the quantities at the same value of α_a before the correction. $\mathbf{P}(\alpha_a)$ and $\underline{\hat{x}}(\alpha_a)$ are called the *a priori* values. The matrix $\mathbf{K}(\alpha_a)$ is sometimes called the *Kalman* weight. The net effect of the Kalman filter is to weight incoming observations depending on the current estimate confidence $\mathbf{P}(\alpha_a)$ and the *residual.* This latter quantity is the difference between the actual and the expected value of the observation, indicated by the term $[\underline{z}(\alpha_a) - \underline{h}(\underline{\hat{x}}(\alpha_a))]$ in the first equation (10.23). Note that the use of this same equation will result in instantaneous changes in the state of the system.

To complete the estimation strategy, we need to take into account the time needed to process an observation acquired by the system. This holds true for any kind of sensor – vision, sonar, laser range-finders, etc. – with more computing time needed to extract the relevant information from complex data. For instance, more time will be needed to process a full camera image than to read a single sonar reading produced by an ultrasonic transducer. As computers get faster this computing time reduces, but it will never be zero. The system does not stop moving while the observations are acquired: doing so would mean that the human teacher needs to pause every time the sensors detected a feature of the environment!

Since α continues to grow while the program is processing the information from the detecting devices, the observations obtained to update the estimates must be transitioned from α_a – the value of α when a sensory sample was acquired – to α_p the value of α when the information has been processed and is available to the Kalman Filter.

So, the corrected version of $\underline{\hat{x}}(\alpha_a \mid \alpha_a)$ becomes:

(10.24) $$\underline{\hat{x}}(\alpha_p \mid \alpha_a) = \underline{\hat{x}}(\alpha_p) + \mathbf{\Phi}(\alpha_p,\alpha_a)\mathbf{K}(\alpha_a)[\underline{z}(\alpha_a) - h(\underline{\hat{x}}(\alpha_a))]$$

The new quantity introduced is denoted by $\mathbf{\Phi}$, and it is called the *state transition matrix.* It is obtained by integrating numerically the equation

(10.25) $$\frac{d\mathbf{\Phi}(\alpha_p,\alpha_a)}{d\alpha} = \underline{f}(\underline{\hat{x}}(\alpha_p), u(\alpha_p))\,\mathbf{\Phi}(\alpha_a,\alpha_0)\,,$$

with initial condition $\mathbf{\Phi}(\alpha_0,\alpha_0) = \mathbf{I}$, the identity matrix. A similar correction needs to be done in the case of the estimation error of the covariance matrix at α_p given the observations up to α_a:

$$\mathbf{P}(\alpha_p \mid \alpha_a) = [\mathbf{I} - \mathbf{\Phi}(\alpha_p \mid \alpha_a)\mathbf{K}(\alpha_a)\mathbf{H}(\hat{\underline{x}}(\alpha_a))]\,\mathbf{P}(\alpha_p)\,. \tag{10.26}$$

To apply our version of the Kalman filter to the moving wheelchair, we first initialize the state of the system (step 1), then keep integrating the equations of motion and the covariance matrix (step 2). Every time we acquire an observation, we need to execute step 3, which corrects our estimates. Otherwise we just keep applying step 2.

We still have not specified how we obtain observations, or how we calculate an observation equation. The discussion up to this point has been general: in order to complete the estimation model we need to indicate $\underline{\boldsymbol{h}}$. This term is known as the observation equation, and it is the mathematical model of a sensing transducer.

Sensors are devices that obtain information from the workspace. Once it has been properly interpreted, this information can be used to make inferences about the environment and produce observations. There are many types of sensors, and each one has its own advantages and drawbacks. Amongst the most commonly used in robotics are active/ passive vision, light detection and ranging (LIDAR), and ultrasound range-sensing (sonar).

In Chapter 5 we discussed the need to use vision as the main sensing paradigm in robotics. The *CPWNS* uses two Costar CV-M50 cameras with mounted Edmund Scientific lenses. These lenses have a focal length *fl* of 8.5 mm and provide a wide view of the scene. As a rule of thumb, perspective effects are significant when a wide-angle lens is used. This allows us to use a simple pinhole camera approach to model our observation equation. In contrast, telephoto lenses create images that approximate an orthographic projection.

The cameras are bolted below the wheelchair seat, and face opposite directions. The left camera captures the right-side view and the right camera the left-side view. This particular setup is meant to take maximum advantage of the wide angle lens of the devices. *Figure 10-10* shows the camera placement (both cameras are behind the coiled wire).

In order to be able to apply the Kalman Filter, we need both the observation equation $\underline{\boldsymbol{h}}$, and the residual, i.e. the term $[\underline{z}(\alpha_a) - \underline{h}(\hat{\underline{x}}(\alpha_a))]$ in equation (10.24). This quantity is the difference between the obtained observation modeled using $\underline{\boldsymbol{h}}$, and the expected observation $\underline{z}$. In the absence of modeling noise $\underline{z} = \underline{\boldsymbol{h}}$, as indicated by equation (10.20).

Figure 10-10. Camera setup

How can we obtain this expected observation?

A possible approach would be to use natural features of the environment, such as corners, wall texture, electrical outlets, furniture, etc. If the position of these features is known with respect to an absolute reference, they can provide sufficient information to obtain $\underline{z}$ once they are detected by a camera. This sounds good in theory, but what about practice?

In practice, this is extremely difficult to do. There are two big problems implied in the last paragraph, and they have proven to be formidable. One is related to map-building, that is, how to create a map of the environment that is accurate and that can be updated (if we cannot account for changes in the workspace, the map is no longer accurate). Over the years several strategies to build reliable maps have been tried. In many of these strategies, the complexity of the environment results in complex models or huge sets of data. The second issue is related to object or feature recognition. We have seen in Chapters 1 and 5 how this problem has vexed engineers and scientists since the early days of Shaky and "hard" Artificial Intelligence. Computers have been proven adept at playing chess, but not at telling the difference between two objects: e.g. a cellular phone and a pair of glasses.

There is a possible, compromised way around these two problems: use a predefined set of cues or beacons. These beacons have a fixed form, which allows the computer to distinguish them quickly in an image. *Figure 10-11* shows two types of cues used in the *CPWNS* project.

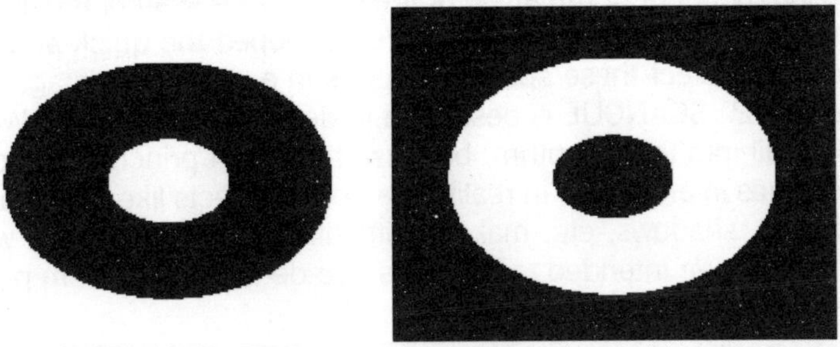

Figure 10-11. Types of cue.

The one with a black ring is called a "black cue" and the one with a white ring is called a "white cue". Consider the image obtained from a wheelchair camera shown in *Figure 10-12*.

Figure 10-12. Camera image.

The computer can detect quickly the black and white cues in the image. The advantage of their shape is that the black and white rings are invariant to perspective deformation. That is, the ratio of black-to-white-to-black regions - in the case of the white cue - is independent of the angle of the camera. A scan of the image can give us the rough position of the cue centers. We developed the quick and simple algorithm *SCANCUE* to detect these special beacons in a grey-scale image, such as the one in *Figure 10-12*. SCANCUE is described in detail in Appendix C. We decided to go deeper in detail into the algorithm, because though in principle it should be easy to detect these cues in an image, in reality it is not so. Effects like cue-shaped objects, lighting conditions, shadows, etc. make it difficult to do it in a robust way. The key word here is *robust*. We intended to have this cue-detection algorithm perform in any environment, not only in the laboratory.

Compare the relatively simple task of finding these predetermined shapes in an image, with the much more ambitious one of recognizing the objects in it. Recognizing a feature of the environment serves as a reference point in approaches like SLAM. This idea could be tried by detecting the vertical lines in the image of *Figure 10-12*, for instance. Each line could be a potential wall corner. This guess is correct in the case of the white shape where the cues are – it is in fact, a wall – but it is mistaken in the case of the grey horizontal rectangle at the center right of the image. This rectangle is a platform in an adjacent room. On top of it, there is a robotic arm, with many vertical lines that could be potential walls. The rectangle at the bottom center of the image, above what looks like a cylinder, is part of the wheelchair. If the system manages to recognize a corner between two walls, like the one at the center left of the image, it could try to estimate its position by finding it in a pre-programmed map. This observation is at most as precise as the map itself. If the system is building the map as it moves, then it has to check if the detected feature has already been stored or if it is new. Moreover, recognizing 3D objects can be a daunting task, since they look very different depending on the angle of the camera, lighting conditions, etc.

Of course, the cue solution is not panacea either. The tradeoff is that it requires setting cues in the environment, with the *CPWNS* working only in that particular workspace. The cues are made of paper, printed in a common desktop printer, and their centers have to be set at 13.5 inches from the floor's surface. This measure corresponds to the center of the lens of the onboard cameras. The positions of the cues have to be hand-measured with respect to a predefined origin point. We would normally choose a corner of the room, with its walls as natural X and Y axes.

We spent many hours setting up cues and measuring distances at the Hines, VA hospital. We would work long hours in the night – not to disturb the patients – to find out in the morning that the cleaning staff had taken them off. Or worse, taken them off and put them back at "more or less" the same position. Linda Fehr, our collaborator at the hospital, can attest how puzzled people were at these unseemly "targets", as they called them.

However, the main problem of the cue approach is the matter of error in measurements. Assuming that the measurement of the cues with respect to a reference point was done as carefully as possible — not always the case, as we found to our chagrin - there will always be a precision error.

If the system calculated its position by triangulation, using two reference points, its performance would depend strongly on the relative exactitude of the cue position measurements. It could have the best possible cameras, the fastest processors and best components, and still perform very poorly. It could maybe traverse a hallway, but probably could not cross a door. A system is only as good as it worst component. In other words, such a system would not be *robust* and would be unviable for a real application.

The estimation part of our approach can represent the errors we have in both our model – using error covariance matrices $\mathbf{P}$ and $\mathbf{Q}$ – and our observation paradigm — using $\mathbf{R}$. However it cannot deal with the extra burden of measurement errors in the reference frame. This is a crippling problem, since the system performance depends strongly on getting reliable observations. How can we deal with this extra source of error?

The beauty of the extension of the teach-repeat approach is that the skewness of this error is always the same. During the teaching episode, the system collects pose data containing errors due to the imprecision of the 'real' location of the cues. When it tries to retrace the taught path, the observations the system collects still have this error, but it is the same as during the teaching episode. The taught positions are with respect *to the reference position of the cues* and not with respect of the absolute frame!

The implication is that we can have errors in the model, observation equations and beacon position, and still have the system do a precise, repeatable and robust performance in a complex, realistic environment. Of course, it is desirable that the position of the cues is as close as possible to their 'real' position in the reference frame. This no longer requires careful, painstaking measurement. In fact, at Hines Hospital our 'instruments' were a couple of wooden rulers and a measurement tape bought at the nearest supermarket.

The problem of setting up the cues could be made easier by having them printed already on an adhesive tape. The spacing between them could be even, to minimize measurement error. A team of workers with no engineering training could set them up in a house in a couple of hours. Once the environment is set, anybody could push the chair and teach it a number of paths.

One last word about the cues: at the hospital and some other venues, people did not like the idea of filling the lower part of their walls with them. We guess we could say the same thing about the esthetics of electrical outlets. Their utility outstripped the misgivings the common person had of "defacing the walls". Still, it is true that the cues look ugly. A possible solution would be to make them invisible. That is, that they could only been seen in a range of the spectrum that the human eye cannot detect, like the infrared frequency. In principle we could print cues in an "infrared ink" and use an infrared lens on the cameras. Another advantage would be that the cameras would *only* detect the cues, essentially erasing any distracting artifact in the image.

Unfortunately, these infrared inks do not work very well. We tried with several types of them, and we could not get reliable detection. Moreover they are not really invisible: they look light green or brown. In the end we decided to stick with our ugly but useful printed black and white cues. Maybe further advances in technology will give us infrared ink that lives up to our expectations.

Returning to our modeling discussion, we now know how to obtain $\underline{z}$, the expected observation: it is the hand-measured position of a cue detected by a camera. We still need to define our observation equation $\underline{\boldsymbol{h}}$. We have mentioned that our video cameras can be modeled as a pinhole camera.

An ideal pinhole camera produces an image that is a perspective projection of a scene. In this model, only the light rays that travel from the scene through the pinhole create an image. The image is resolved in the *image plane*. Refer to *Figure 10-13*.

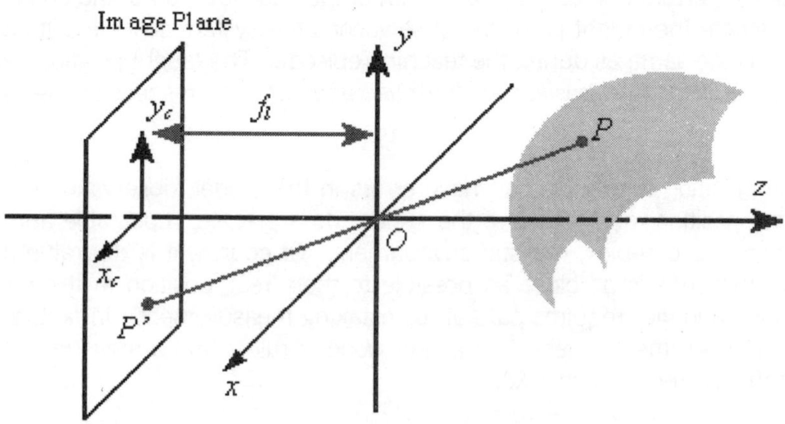

Figure 10-13. Pinhole Camera Model.

In the figure x, y, z are the world or scene coordinates with origin at pinhole O. x_c and y_c are the coordinates of the image plane. P is a point of an object or scene and P' is its projection O in the image plane. In the case of perspective or central projection, the projection of a three-dimensional scene or object onto a plane using straight lines passes through a single point, called *focal point* or *center of projection.* The pinhole O coincides with the focal point. The *focal length* is the distance of the image plane from the focal point, and it is represented by f_l in the figure. From the model, the following relationships between the points in the scene and the image plane can be deduced:

$$x_c = f_l \frac{x}{z}$$

(10.27)

$$y_c = f_l \frac{y}{z}$$

These relationships define the perspective projection and will be useful to calculate the observation equation of our model.

Figure 10-14 is a representation of the pinhole camera model in the context of the wheelchair vehicle. Point ***A*** is the midsection of the axle that joins the two wheels, and is the point that is used to determine the position of the vehicle in the global coordinate system. In the figure, focal point O has global coordinates X_f and Y_f.

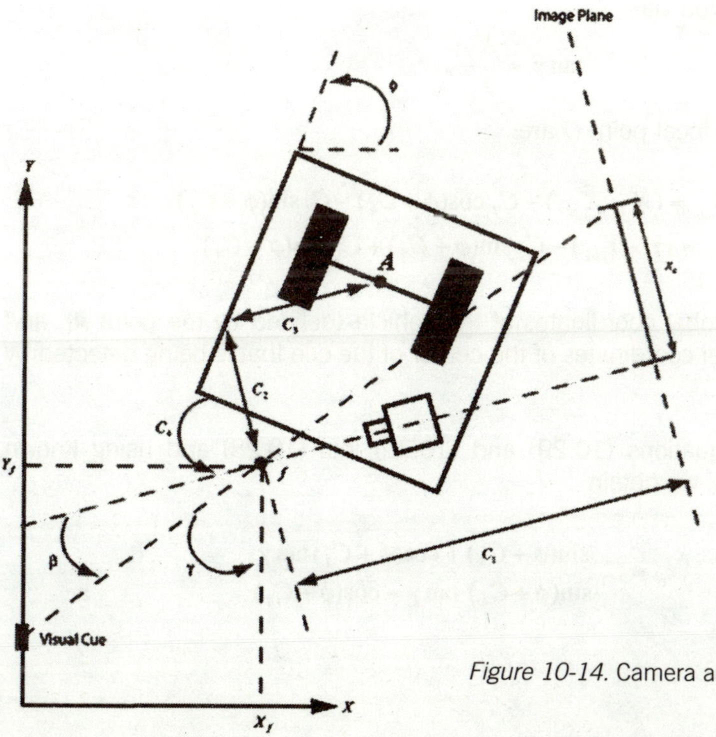

Figure 10-14. Camera and Vehicle Parameters.

A series of calibration parameters, C_1 - C_4, describe physical characteristics for each camera, and are assumed to be held fixed. C_1 is the focal length of the camera measured in pixels, C_2 and C_3 are the coordinates of the focal point with respect to the point of interest ***A***, and C_4 is the orientation of the camera relative to the vehicle base.

These parameters can be estimated by a nonlinear procedure called Marquart's method. Eric T. Baumgartner originally obtained the values for the cameras of an earlier prototype, and Brian Reichenberger did the necessary calibration for the most recent version.

In the model described in *Figure 10-14*, the horizontal position x_c of a cue in the image plane of the camera is given by:

(10.28) $$x_c = C_1 \tan \beta$$

with

(10.29) $$\beta = \pi - (\phi + C_4 + \gamma).$$

Here ϕ is the orientation of the vehicle with respect to the horizontal axis. The angle γ can be calculated via:

(10.30) $$\tan \gamma = \frac{X_f}{Y_f}.$$

The coordinates of the focal point ***O*** are:

(10.31) $$\begin{aligned} X_f &= (X - X_{cue}) - C_2 \cos(\phi + C_4) - C_3 \sin(\phi + C_4) \\ Y_f &= (Y - Y_{cue}) - C_2 \sin(\phi + C_4) + C_3 \cos(\phi + C_4) \end{aligned}$$

where X, Y are the global coordinates of the vehicle (defined by the point ***A***), and X_{cue}, Y_{cue} are the global coordinates of the center of the cue that is being detected by the camera.

By substituting equations (10.29) and (10.30) into (10.28) and using known trigonometric identities, we obtain

(10.32) $$x_c = C_1 \frac{\sin(\phi + C_4) + \cos(\phi + C_4) \tan \gamma}{\sin(\phi + C_4) \tan \gamma - \cos(\phi + C_4)}.$$

Finally, substituting (10.31) in (10.32) we can express x_c in terms of the known parameters C_1 - C_4 and the position and orientation of the vehicle, given by X, Y, ϕ,

$$x_c = C_1 \frac{(X - X_{cue})\cos(\phi + C_4) + (Y - Y_{cue})\sin(\phi + C_4) - C_2}{(X - X_{cue})\sin(\phi + C_4) - (Y - Y_{cue})\cos(\phi + C_4) - C_3} \equiv h(\underline{x}; X_{cue}, Y_{cue}, \underline{C}) \tag{10.33}$$

Here $\underline{x} = (X, Y, \phi)^T$ is the state vector, $\underline{C} = (C_1, C_2, C_3, C_4)^T$ is the vector of camera parameters, and $\boldsymbol{h}$ is the observation equation.

We finally have all the elements needed for a teaching episode. Before we get into the actual data of an example, let us take a look at our chosen environment.

A special testing area has been arranged at Dexterity, Vision and Control Laboratory at the University of Notre Dame. The purpose of the current setup is to have a complex workspace to test the navigational capabilities of the *CPWNS*. *Figure 10-15* shows a schematic of the current workspace.

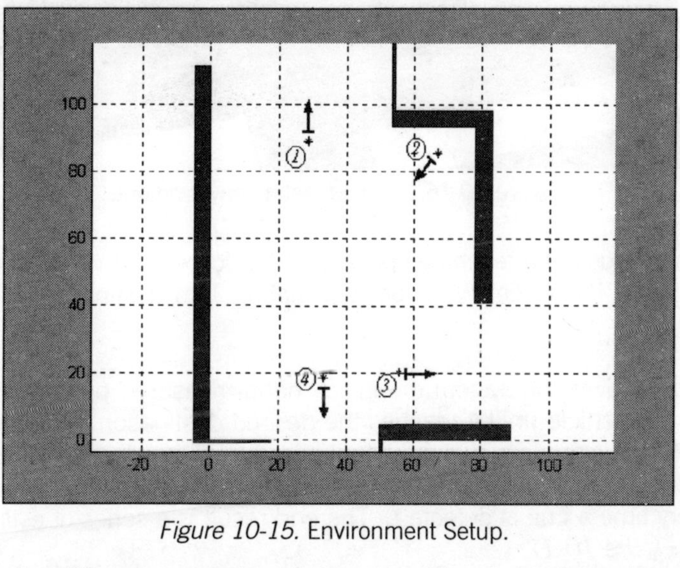

Figure 10-15. Environment Setup.

The figure is a 2D representation of the room. The plane defined by the **X** and **Y** axes coincides with the plane of the floor. The three polygons in the center represent three makeshift walls made of plaster. The numbers corresponds to four possible predetermined destinations for the *CPWNS*, and the arrows point to the forward part of the vehicle at the indicated position. The localizations for the points on the floor are:

- Destination 1 has coordinates **X** = 81.5 in., **Y** = 160.5 in.
- Destination 2 has coordinates **X** = 119.0 in., **Y** = 149.5 in.

- Destination 3 has coordinates **X** = 109.5 in., **Y** = 93.0 in.
- Destination 4 has coordinates **X** = 87.0 in., **Y** = 77.5 in.

Figure 10-16 shows what the testing environment looks like. Its tight space allow us to test the capacity of the wheelchair to reach any of our four preset destinations.

Figure 10-16. Photo of testing environment.

Let us take a look at a teaching episode. What follows are video stills taken while a student pushed the wheelchair from destination 1 to destination 4. *(See oposite page)*

The student gives the system the initial, hand-measured position. Then, physically pushes the vehicle until it reaches the desired destination. While the system is being pushed, the encoders give rotational information needed for the dead-reckoning integration, while the Kalman Filter incorporates the information obtained from the cameras every time a cue is detected. The end result is a series of estimated poses, as shown in *Figure 10-17*.

The triangle denotes the starting point of the trajectory, and the square the end. The stars represent a change in the state equations. In our case, the wheelchair was first pushed backwards, then pivoted counterclockwise, followed a stretch moving backwards, and finally moved forward. We are able to detect these changes by monitoring the values of u and u^* as we previously mentioned. The points in *Figure 10-17* represent the path followed by the midpoint of the axle of the wheels while the

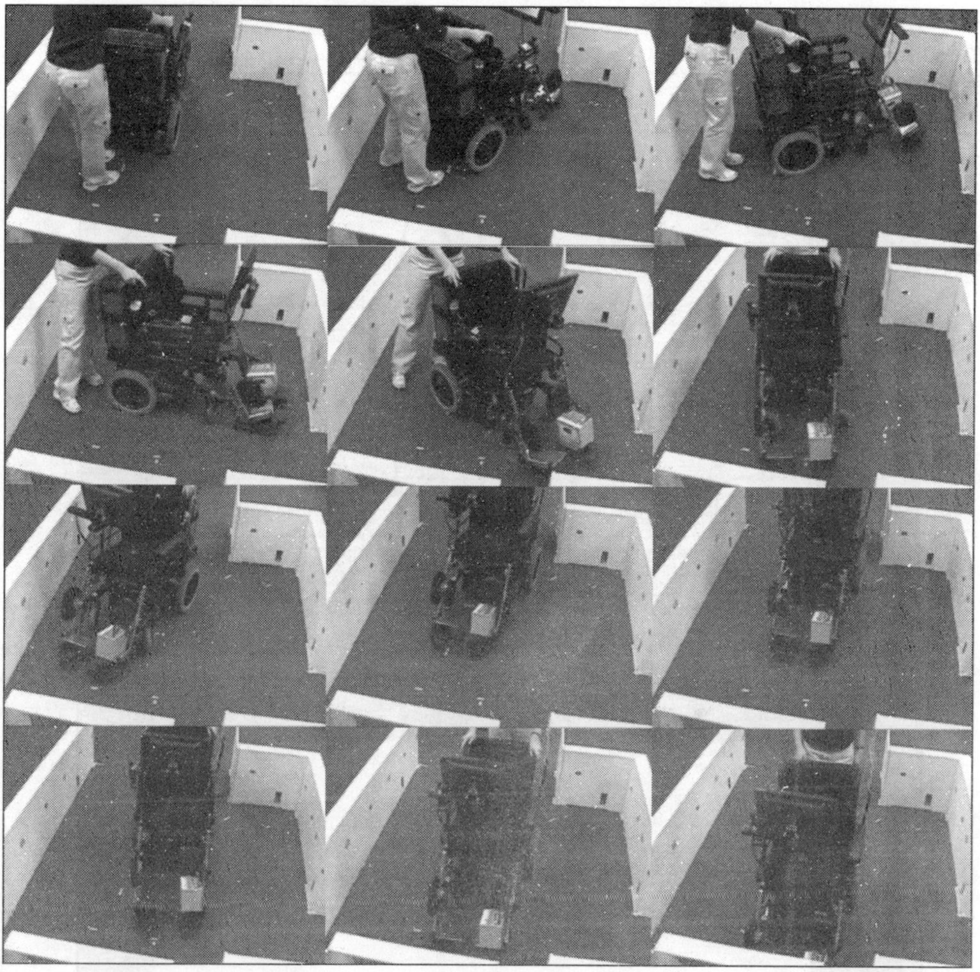

wheelchair was being pushed. They look contiguous, but appearances are deceiving. Let us take a closer look at the end of the trajectory in *Figure 10-18*.

First notice how the series of points have small regular jumps. Then there is a large gap at the end, at which point the wheelchair is stopping and there is a lot of switching between movements. These are symbolized by the stars in the figure (they seem to be three, but there are about 40!). The mode switching is noise, due to small encoder readings when the wheelchair is essentially stopped. The large jump is due to large Kalman filter correction at the end of the teaching episode.

It would be hard for the wheelchair to try to follow a series of disconnected points, so we need to store and use this information in a more effective way. We will in fact

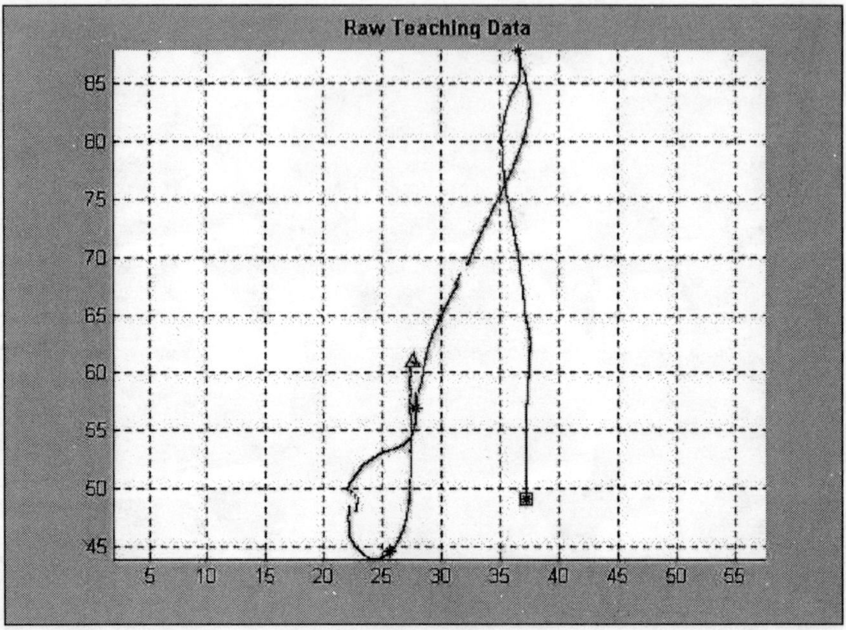

Figure 10-17. Raw Teaching Data.

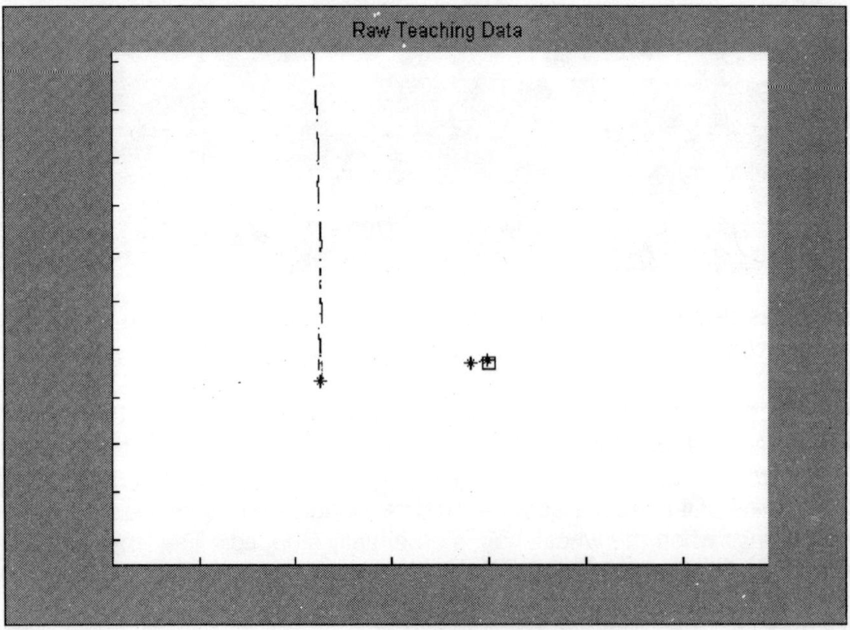

Figure 10-18. Raw Data (Close-up).

convert these points into a series of line segments. This is called *post processing*, and it is a necessary step between the teaching and tracking episodes.

Once the teaching episode is completed, the series of points that the program generated has to be transformed into line segments. This process is done offline, so that this burdensome task does not take place either in the teaching or the tracking programs. The postprocessor is responsible for creating a reference path that the tracking program can follow. The reference path is represented as a series of line segments. In order for a path to be viable, its component segments are created in compliance with certain rules.

A great advantage of using an offline path-generating algorithm is that the segments are not created in real-time. The teaching program simply deals with the estimation of the state, allowing more integration steps and a relatively large number of observations. A more accurate representation of the path is achieved by getting more observations per unit of distance traveled. All state estimate information is saved to disk, an operation that requires little computer time. Then the offline path-generating algorithm disposes of all the computer resources to generate a line segment representation of the taught trajectory.

Performing this process offline offers other advantages. They have to do with the stability of the tracking episode, and other control concerns such as noise and discontinuities in the teach data. Dealing with them in detail would require space and some knowledge of control theory which are outside of the scope of this chapter.

Hence we assume that we can obtain a variable series of line segments from the teaching data. Each of these segments has a mode of movement assigned to it, so that the tracking program knows which set of state equations u and α to use. The result of the postprocessor on the dataset of *Figure 10-18* is shown in *Figure 10-19*. The squares indicate the boundaries of the line segments, and the stars the changes of mode. These line segments are the *reference* path that the wheelchair will track.

Let us take a closer look at s and s^* our definitions of the distance traveled. This analysis will provide us with a control strategy that we can use for the tracking of the taught path. First we will consider the case where both wheels move in the same direction, forwards or backwards.

After the postprocessor program has reduced the teaching sequence to a series of line segments, we parameterize each segment by a normalized s ranging from 0 to 1. At the starting end of the interval ($s = 0$) estimates from the teaching are assigned to create the reference path. At the terminal end ($s = 1$) the same occurs. This line segment is the trajectory that the tracking program will attempt to follow. The parameterization is shown in *Figure 10-20*.

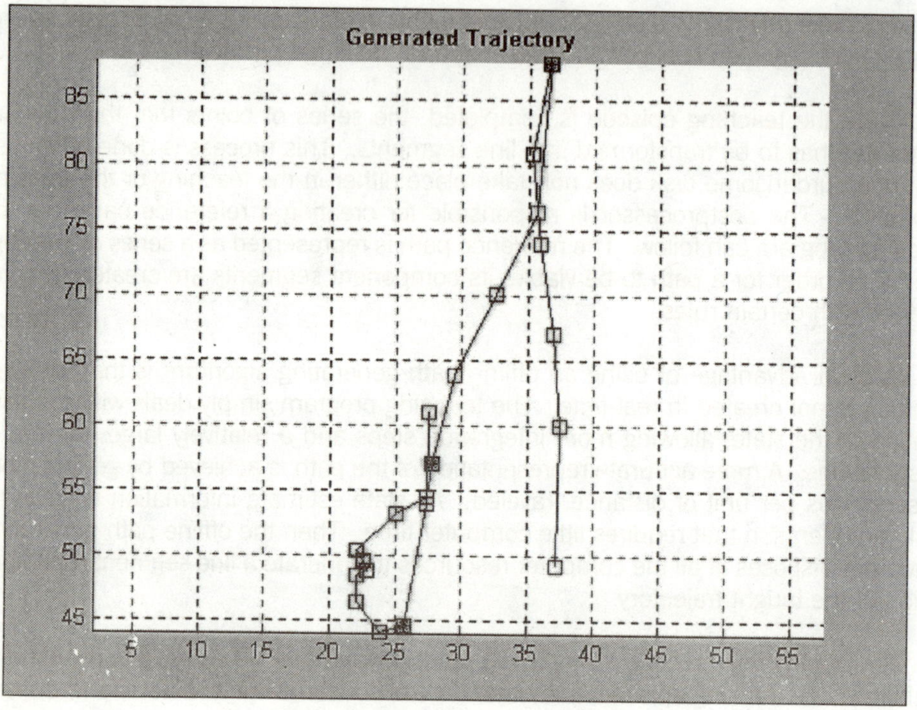

Figure 10-19. Trajectory segments.

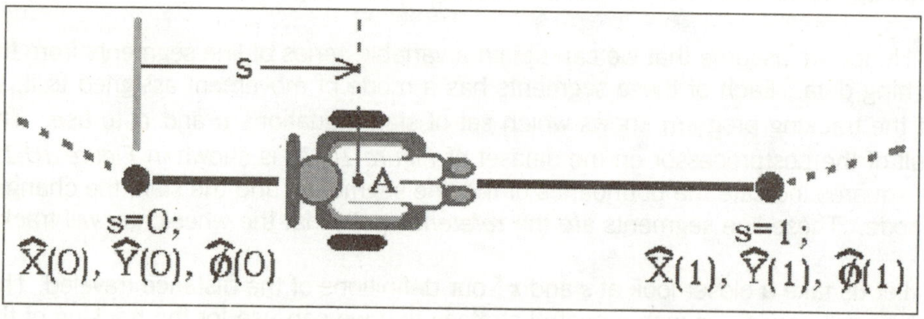

Figure 10-20. Parameterization of reference path.

We can then define the reference values of a segment of the trajectory X_{ref}, Y_{ref}, and ϕ_{ref}, as functions of the distance traveled s

(10.34)

$$X_{ref}(s) = X_{ref}(0) + s[X_{ref}(1) - X_{ref}(0)]$$
$$Y_{ref}(s) = Y_{ref}(0) + s[Y_{ref}(1) - Y_{ref}(0)]$$
$$\phi_{ref}(s) = \phi_{ref}(0) + s[\phi_{ref}(1) - \phi_{ref}(0)]$$

where the points $(X_{ref}(0), Y_{ref}(0), \phi_{ref}(0))$ and $(X_{ref}(1), Y_{ref}(1), \phi_{ref}(1))$ are the end-points that the postprocessor calculated for each segment of the trajectory. Since we defined s to range from 0 to 1, the tracking program will know when to switch to the next segment. That is, a segment is completed when $s \geq 1$. *Figure 10-21* shows the reference point corresponding to a wheelchair pose $\hat{\underline{x}}(\alpha) = (\hat{X}(\alpha)\ \hat{Y}(\alpha)\ \hat{\phi}(\alpha))^T$. In the figure we use the notation $\underline{x}_r(s) = (X_{ref}(s)\ Y_{ref}(s)\ \phi_{ref}(s))^T$.

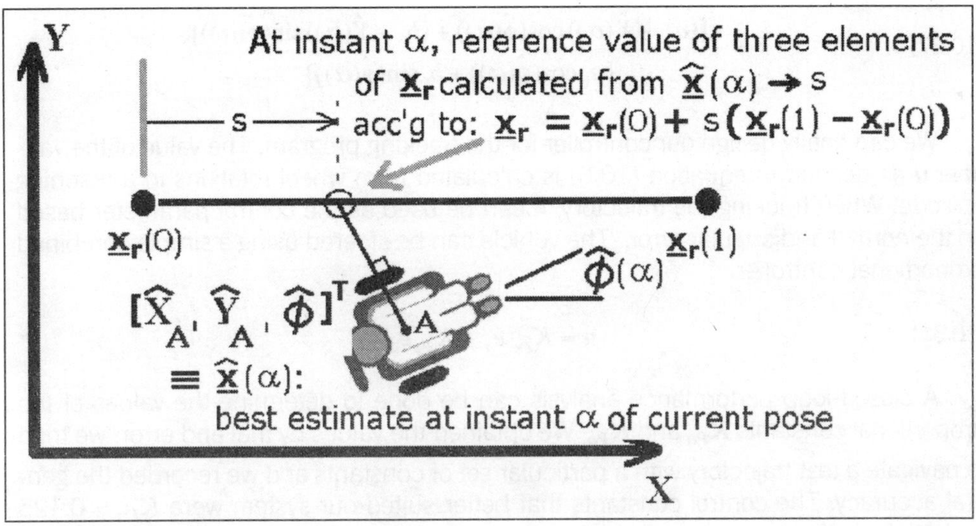

Figure 10-21. Identification of reference point for a given α.

Let us introduce three quantities that measure the difference between the estimated position of the chair and the position where the vehicle should be at a certain juncture α. These variables will help design a viable tracking control for the wheelchair, and are defined according to:

$$
\begin{aligned}
e_n(\alpha) &= -[X_{ref}(\alpha) - \hat{X}(\alpha)]\sin(\hat{\phi}(\alpha)) + [Y_{ref}(\alpha) - \hat{Y}(\alpha)]\cos(\hat{\phi}(\alpha)) \\
e_\phi(\alpha) &= \phi_{ref}(\alpha) - \hat{\phi}(\alpha) \qquad (10.35) \\
e_t(\alpha) &= [X_{ref}(\alpha) - \hat{X}(\alpha)]\cos(\hat{\phi}(\alpha)) + [Y_{ref}(\alpha) - \hat{Y}(\alpha)]\sin(\hat{\phi}(\alpha))
\end{aligned}
$$

Here e_n is the error normal to the trajectory, e_ϕ is the difference in the angle between the desired and the actual in plane orientation, and e_t is the error in the tangential direction. We can see that the magnitude of the error in position of the wheelchair at a certain juncture is

$$|e_{pos}(\alpha)| = \sqrt{e_n^2(\alpha) + e_t^2(\alpha)}\,. \qquad (10.36)$$

In *Figure 10-21*, the vector that connects point ***A*** with the reference segment is the error in position. The error in position has only a normal component, since the wheelchair cannot move sideways while both wheels move in the same direction. Due to that physical constraint, we can assume that the tangential error is zero. As it happens, by using $e_t = 0$ and equations (10.34) we can calculate the distance traveled s since all other quantities are known. Then

(10.37) $$s = \frac{-[(a_0 - \hat{X}(\alpha))\cos(\hat{\phi}(\alpha)) + (b_0 - \hat{Y}(\alpha))\sin(\hat{\phi}(\alpha))]}{[a_1\cos(\hat{\phi}(\alpha)) + b_1\sin(\hat{\phi}(\alpha))]}.$$

We can finally design our controller for the tracking program. The value of the variable u as defined in equation (10.5) is calculated from wheel rotations in a teaching episode. When tracking the trajectory, it can be used as the control parameter based on the normal and angular error. The vehicle can be steered using a simple combined proportional controller

(10.38) $$u = K_{Pn}\,e_n + K_{P\phi}e_\phi.$$

A closed-loop performance analysis can be done to determine the values of the proportional constants K_{Pn} and $K_{P\phi}$. We obtained the values by trial and error: we tried to navigate a test trajectory with a particular set of constants and we recorded the general accuracy. The control constants that better suited our system were $K_{Pn} = 0.125$ and $K_{P\phi} = 3.75$.

In an analogous way we can design the control when the wheelchair is doing a rotational maneuver. In this case we use the angular distance traveled s^*, rather than s.

The values of the reference path at a certain juncture are calculated in a similar way to equations (10.34):

(10.39) $$\begin{aligned} X_{ref}(s^*) &= X_{ref}(0) + s^*[X_{ref}(1) - X_{ref}(0)] \\ Y_{ref}(s^*) &= Y_{ref}(0) + s^*[Y_{ref}(1) - Y_{ref}(0)] \\ \phi_{ref}(s^*) &= \phi_{ref}(0) + s^*[\phi_{ref}(1) - \phi_{ref}(0)] \end{aligned}$$

where s^*, takes values from 0 to 1. The quantity s^* depends only on ϕ, and we can calculate it directly,

(10.40) $$s^* = \frac{\hat{\phi}(\alpha^*) - \phi_{ref}(0)}{\phi_{ref}(1) - \phi_{ref}(0)}$$

since the value of $\hat{\phi}(\alpha^*)$ is known at all times. Observe *Figure 10-22.*

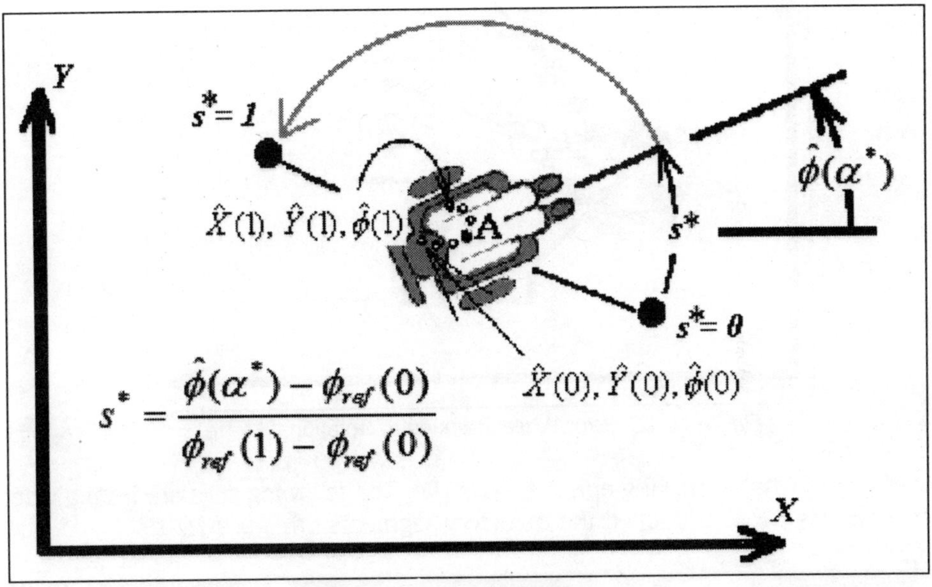

Figure 10-22. Definition of s*

Since the values of X_{ref} and Y_{ref} can be calculated from s^*, it is possible to calculate the ongoing position error while trying to track a trajectory. Then, e_n, e_ϕ and e_t can be calculated using equations (10.35).

The strategy for computing u^* is to actuate the wheels in order to correct the current translational error to the extent to which is possible. This is e indicated in *Figure 10-23.* Since instantaneous progress in translation can only occur in the direction of the chair, the component e_t (the translational error) is the meaningful one in this context.

The proposed control law is

(10. 41) $$u^* = K_p\, e_t\,.$$

Over the rotational course of the maneuver, the resulting commanded angular velocity for the two wheels should tend to diminish current translational error, while undergoing the required rotation. As in the translational case, the value of the proportionality constant was obtained by extensive testing and was set at $K_P = 0.3$.

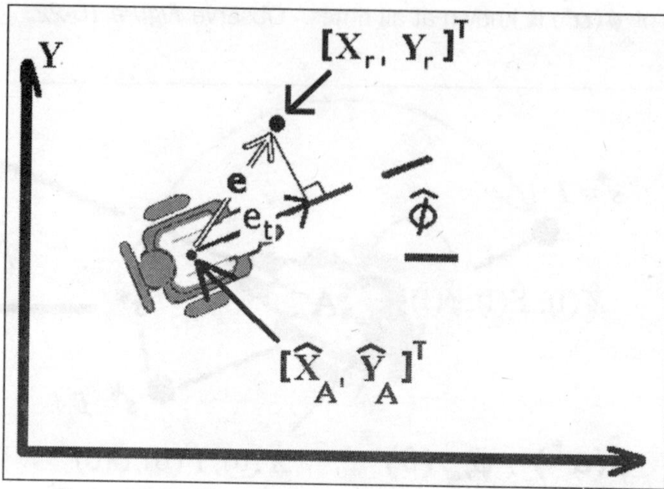

Figure 10-23. Error While Tracking in Rotational Mode.

Let us see what a tracking episode looks like. The following stills are from a video taken while the *CPWNS* tracked the trajectory segments of *Figure 10-19*.

The wheelchair seems to follow closely the series of poses previously taught. How accurate is the tracking? The tracking program can write its estimates to file, allowing us to plot the results, superposing them with the line segments that the system is trying to follow. The next images show the tracking data.

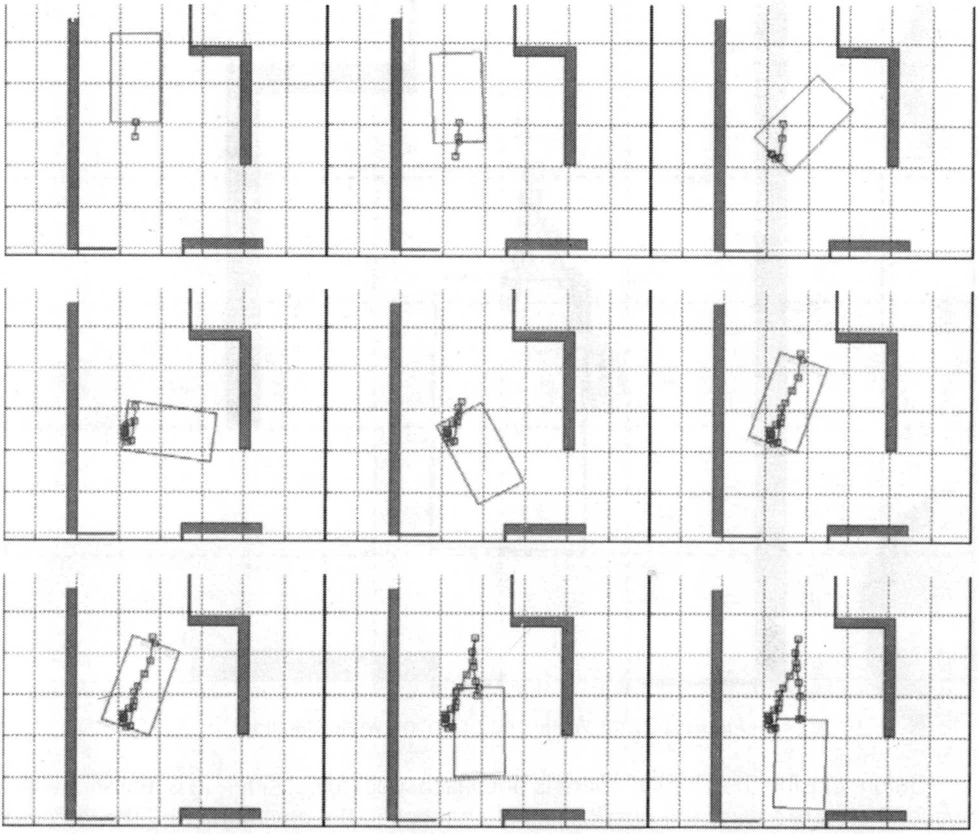

The polygons represent the walls in the environment. The wheelchair is represented as a wireframe rectangle and the trajectory is represented as a series of line segments. These "stills" give us an idea of the precision of the maneuver. *Figure 10-24* shows an enlargement of the one particular position of the wheelchair.

The small circle is the back of the wheelchair and is 6.75 inches behind the midpoint of the axle – what we called point ***A***. This is the point that the program actually monitors, instead of ***A***, simply because it makes measurement easier.

As we see from the stills and in *Figure 10-24*, the tracking follows closely the taught trajectory. In this example, the position error is always less than one inch and the angular error is less than 8 degrees. Higher precisions are possible. As we mentioned before, the idea is that the vehicle behaves as if it was on a rail, and that it is as safe.

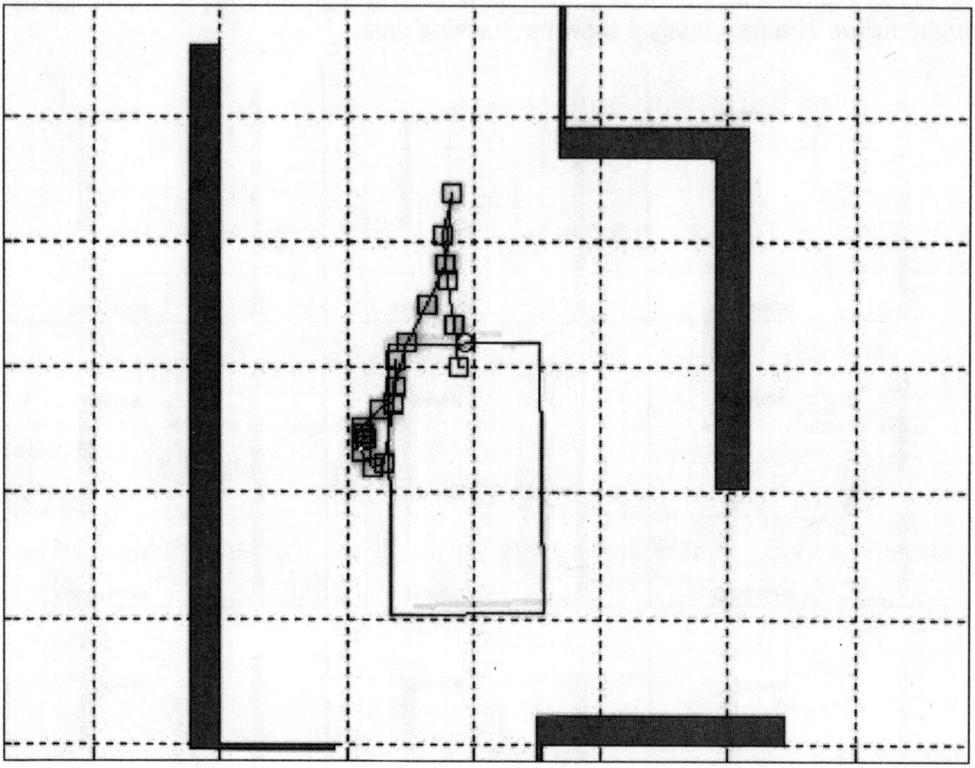

Figure 10-24. Wheelchair position while tracking.

Before closing the chapter, there is one last aspect to consider: in a realistic workspace, there are always the possibilities of changes in the environment. These could be an object blocking the path that was not there previously, a rearrangement of the furniture in a room after the teaching episode, a trajectory that crosses a doorway that has to take into account that the door might be closed, or a path that waits for an elevator to arrive, just to name a few examples.

A moving vehicle has to be able to deal with such eventualities. All the situations described above have in common the necessity of obtaining information to detect unpredicted changes in the environment. This information could then be used to prevent collisions or change the course.

So our first decision is to choose a device to detect range or distance information. This is an important consideration: each device has its own limitations. What is critical is that these limitations are directly related with the physical act of sensing.

Historically, ultrasound range-sensing or sonar has been used since the mid-eighties to generate simple range data. It was the introduction of the Polaroid range sensor as a focusing aid for cameras that started a trend to create in-door maps using mobile robots. There are however several disadvantages to the use of sonar. These limitations have to do with the physics of sound and the characteristics of materials and are explained in detail further below.

In terms of alternatives to sonar, the main contender is light detection and ranging, otherwise known as *LIDAR*. These devices consist of a transmitter that illuminates a target with a collimated beam, usually of infrared light, and a receiver that detects the reflected beam. They can then calculate the elapsed time between emission and detection to generate a range estimate of the distance to a feature of the environment. These are the devices that come closest to portraying a ray-trace scanner behavior, an ideal sensor that we will describe shortly.

LIDAR sensors have several disadvantages. They require a precise optical scanning system and a sensitive, low noise electronic design. LIDAR systems used to be physically large and prohibitively expensive, with non-safe laser transmitters. Recently they have become more manageable in size and cost, and researchers in robotics have started using them. However, they are still by far more expensive than sonar sensors, and they require a special interface to connect to a computer.

As we started to develop a scheme to detect obstacles for the *CPWNS*, we had to use ultrasound sensors. The original reasons were more related to budget and ease of use than anything else: we simply could not afford fancy stuff like LIDAR. In the end, these conditions worked in our favor. We were able to develop a strategy for teach-repeat that takes advantage of the coarse representations that sonar generates.

Before we get to the strategy, it will be useful to understand how ultrasound sensors work. In general, ultrasonic range finders come packaged as a single transducer that works as an emitter-receiver. When the device is triggered, it emits and acoustic wave. Then it switches its mode to receiver, in order to detect the reflected pulse after it has bounced from an object. This simple principle of detection is called time-of-flight detection or TOF. Tha fact that the sensor has to be switched off while emitting the wave and during a recovery time, imposes a physical limitation on the minimum distance that can be detected.

When the echo returns to the transducer after a certain time T, the estimated range r is

$$r = \frac{vT}{2}, \tag{10.42}$$

where v is the speed of sound in air.

A *ray-trace scanner* is an ideal sensor that is pencil-thin in its width and has no specular or diffuse effects. Sonar has been a great source of frustration for researchers that were looking for ray-trace-like data in ultrasound readings. Ultrasound range sensing has two big shortcomings with respect to this imaginary device: beam width and specularity.

To get an idea of how bad is this distortion, compare the plots of ray-trace scanner simulated data and actual sonar readings in *Figure 10-25.*

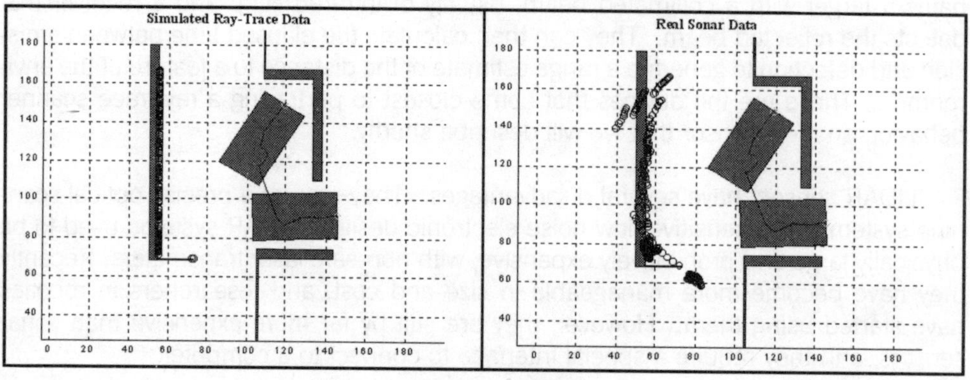

Figure 10-25. Simulated and real sonar data.

In *Figure 10-25* the black circles indicate the sonar readings. The two rectangles indicate the initial and final position of the vehicle. The polygons are representations of the walls of the environment.

In the left figure our ideal scanner has created an accurate representation of a wall. In contrast, in the right figure the wall is detected at different distances, depending on the orientation of the vehicle.

The acoustic impedance of air is quite low, while the typical impedance of solid objects is much larger. The effect is that objects behave like acoustic mirrors. As the sound source diverges from the orthogonal direction to the surface, the energy from the returning echo is lessened. The diminished wave in turn produces a distorted reading of the range information, since objects seem to be further away than they really are. On the right side of *Figure 10-25* however, some of the objects appear closer than they really are.

This phenomenon is related to the other big limitation of sonar, the beam width. Ultrasound is normally not emitted as a focused beam. This results in readings that register the closest feature on an object that gets detected by the spreading wave. In other words, the first signal above the threshold in the conical beam produces the reading. Brown and Kuc described this specular phenomenon. Observe *Figure 10-26.*

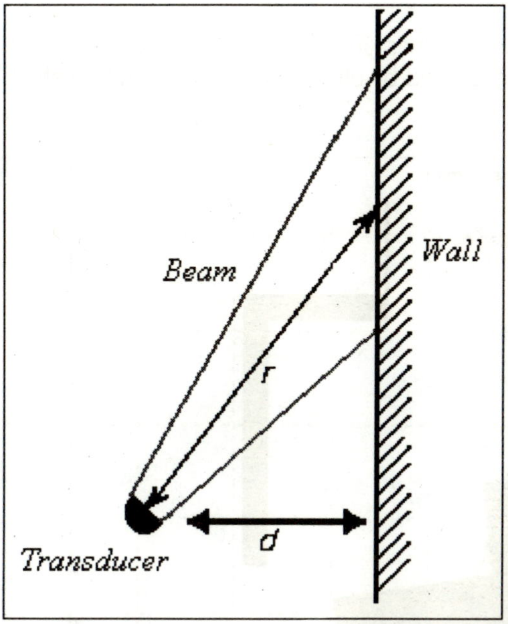

Figure 10-26. Detected distance to a wall.

While trying to organize the received data, if the distance detected is assigned to the point at the center of the beam, the object will appear at the smaller distance d instead of the true distance r. These limitations have made it difficult to create accurate representations of an environment using sonar data.

We will now briefly review an approach proposed by Kuc and Siegel, and later extended by Leonard and Durrant-Whyte. This method tries to make inferences about the workspace taking into account the problematic characteristics of ultrasound range sensing.

As it can be seen on the right side of *Figure 10-26*, the effect due to the beam width creates a series of curved surfaces along walls. This phenomenon is a consequence of the fact that the distance from the wall remains constant, no matter what is the angle of incidence of the sound cone (or the sensor with respect to the wall). These curves are typical of sonar readings.

The curved surfaces were referred to as *regions of constant depth* (RCDs) in Leonard's work. Kuc, Leonard and Siegel suggested that correct sonar interpretation depends on extracting the RCDs from raw data. The work of Kuc and Siegel was to characterize the different surfaces and their matching RCDs.

Let us take a quick look to some of their conclusions:

The response of a wall to ultrasound is a series of arcs at the range of the closest return. *Figure 10-27* shows the real data captured by ultrasound close to such a structure.

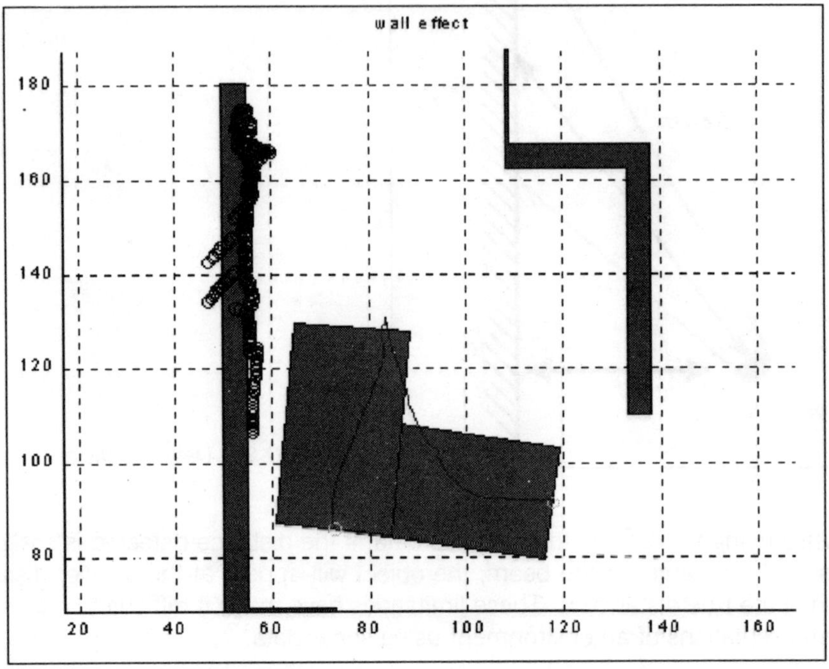

Figure 10-27. Sonar behavior near a wall.

Corners and edges have a particular behavior. Observe *Figure 10-28*: the path of a sonic wave is shown bouncing on a corner.

Any wave that hits a corner wall produces a reading equal to $d_1 + d_2 + d_3$, where each d represents a distance traveled in the plot. If the corner is 90 degrees - as it commonly is – then $d_1 + d_2 + d_3 = l$ for all α, where l is the perpendicular distance to the corner from the transducer plane, and α is the angle of the sensor with respect to the horizontal. This implies that the inclination of the sensor is irrelevant, and that the resulting RCD is the same as the one generated by a wall.

Corners and edges behave as RCDs that vary with orientation depending on the angle of the emitting device. Consider *Figure 10-29*, generated with real sonar data.

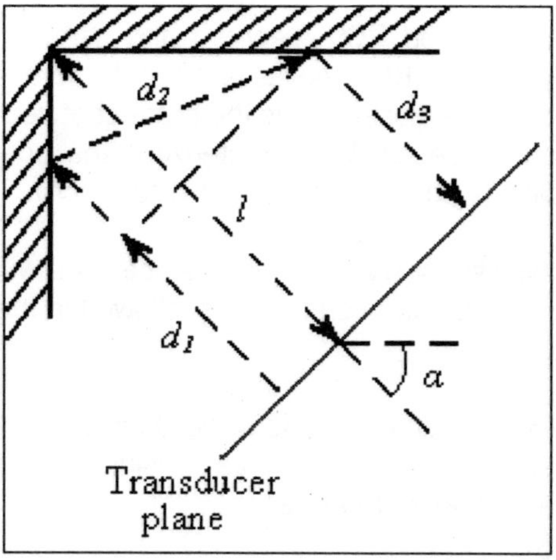

Figure 10-28. Sound trajectory due to a corner.

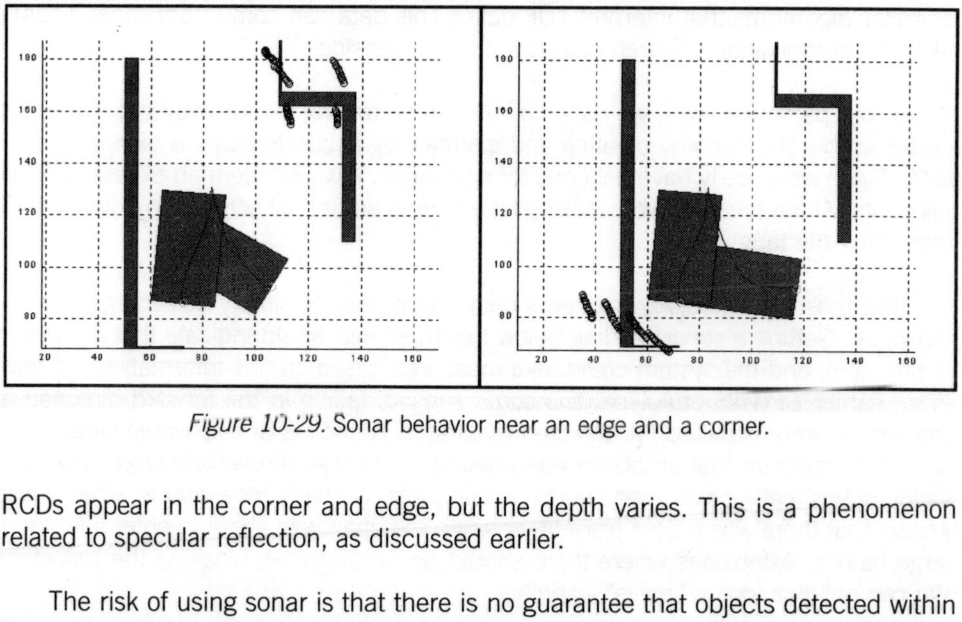

Figure 10-29. Sonar behavior near an edge and a corner.

RCDs appear in the corner and edge, but the depth varies. This is a phenomenon related to specular reflection, as discussed earlier.

The risk of using sonar is that there is no guarantee that objects detected within the cone correspond to an object that is in front of the plane of the transducer while using simple TOF sensing.

Leonard and Durrant-Whyte used sonar as the main sensor for navigation on a mobile robot. They modeled the environment *a priori*, in a hand-measured map that they stored in an onboard computer. They had a simple path-planning algorithm and used the Extended Kalman Filter to do the tracking. The observations were obtained by TOF sonar readings, and the matching of detected RCD with an *a priori* modeled RCD gave an estimate of the position.

To be able to distinguish the origin of a detected RCD, Leonard and Durrant-Whyte followed the behavior of the artifact. Typically, a RCD moves tangentially across a wall and rotates around a corner. All other data are discarded as the product of second order reflections or of artifacts that cannot be easily characterized (the only exception was a cylindrical object, but that case is not relevant for our discussion).

Some of their ideas about RCDs and the way sonar data can be characterized will be important for the following discussion and for the approach that was followed in the *CPWNS* project.

The crucial difference between the *CPWNS* and other robotic wheelchair projects that use sonar — like *Rolland* or *VAHM*, for example — is that range sensing is used *only* for obstacle detection and not as the main means of navigation.

Many recent efforts have tried to build accurate maps of the environment using complex algorithms that interpret TOF data. This data can come from sonar, LIDAR, etc. or a combination of different kinds of range-sensing.

In our particular case, we have no interest in creating maps. We have discussed earlier in this chapter why creating and maintaining accurate maps is extremely difficult. Since we already have beacons for navigation, it would seem an excessive complication. At the time, we were looking for a simple but robust scheme to detect obstacles along the taught path.

The most straightforward scheme could have been to simply designing a threshold value. Getting a sonar reading below this threshold would indicate that a collision is probable, and the system could take measures based on this information. In fact, in an earlier *CPWNS* prototype, two sonar sensors facing in the forward direction of the vehicle were installed. A detected reading of 10 inches or less would send a signal to the program that an object was present, and make the vehicle stop. The activation of this feature was used only in certain parts of a path; for example, when it was known that there was a door present; or when the chair was about to enter a lift; or in large hallway extensions where there should be no obstacles blocking the trajectory. We can call this idea a "naïve" approach.

The problem is that this naïve scheme is not feasible in complex environments - say, a living room or a kitchen- where maneuvers can take the vehicle very close to walls or furniture. In the path shown in *Figure 10-19*, for example, the vehicle can come as close as 3 inches or less from one of the walls. Clearly, this situation could not be negotiated with the approach discussed in the earlier paragraph.

The basis of what became our proposed scheme takes advantage of the two steps used in teach-repeat. First the system collects ultrasound samples during the teaching episode of a certain trajectory. Then this information — after being compressed in a useful form — can be compared with the range readings detected during tracking. Kuc and Siegel showed that the sonar readings given by structures create consistent artifacts — the mentioned RCDs.

Our approach uses the idea of defining regions that bound the sonar readings during the teaching episode. Then, while tracking, the ultrasound readings are compared to them. If readings occur outside of this designated boundary, they would mean that the environment is not the same as when the teaching was performed. In this case, a warning or an evasive procedure can be executed, or the vehicle may be brought to rest. Before we get into the specifics of the approach, let us briefly describe the ultrasound detection hardware of the system.

The ultrasonic range-sensing hardware of the *CPWNS* consists of the sensors, two supporting metal plates, a microcontroller and serial connection to the main CPU. Eight ultrasound sensors are attached to two steel plates, four for each plate. These rigid plates are bolted below the footrests of the wheelchair. Each transducer is set at a different angle from the center axis of the vehicle. This arrangement is meant to increase the possibility that at least one sensor will get a reading close to the perpendicular plane of a possible object or a wall. The sensors are at angles of 81.7, 53.7, 36.3 and 8.3 degrees on the left side; and -81.7, -53.7, -36.3 and -8.3 degrees on the right side. Angles are measured with respect to the axis on the longitudinal direction of the vehicle. *Figure 10-30* shows a photograph of the sonar transducers on the right steel platform.

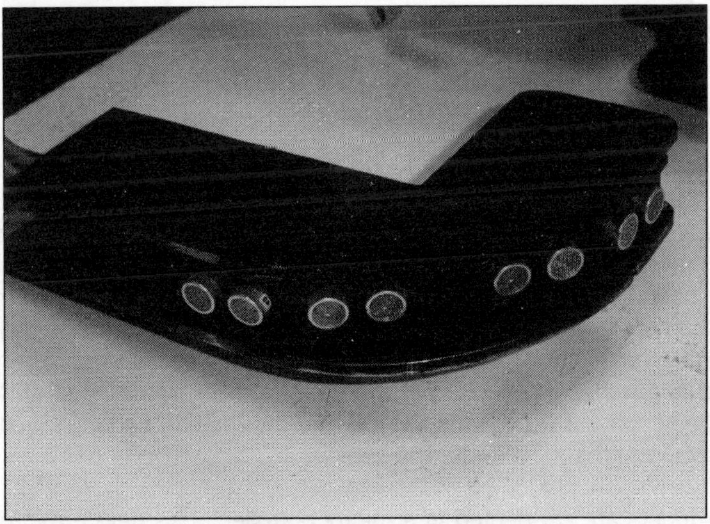

Figure 10-30. Sensor platform.

Figure 10-31 shows the schematic of the right steel plate, with the dimensions in inches.

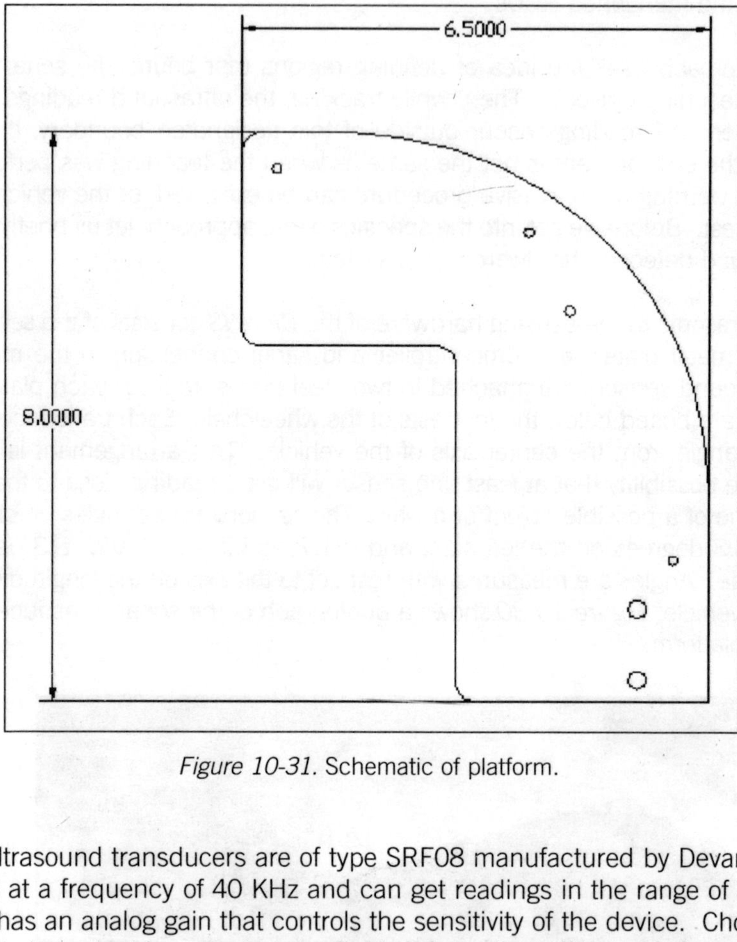

Figure 10-31. Schematic of platform.

The ultrasound transducers are of type SRF08 manufactured by Devantech Ltd. They work at a frequency of 40 KHz and can get readings in the range of 3 cm to 6 m. It also has an analog gain that controls the sensitivity of the device. Choosing the gain is quite important in order to get reliable information. If the gain is too high, the sensors get a bouncing signal from the floor, or get multiple reflections or specularities. This phenomenon occurs when the part of the beam that strikes a wall does not produce an echo that is strong enough to be detected by the device. Then, the first echo detected is a multiple reflection of the beam striking some other surfaces and returning to the sensor. If the gain is too low, the information becomes quite sparse. It took us sevaral months to calibrate the sensors.

The transducers are connected to a microcontroller. This device microcontroller measures the time elapsed between the sound emission and the detected echo of each sonar device, generating the TOF measurement. After the reading is converted to a distance measurement, the information is then sent to the main onboard computer.

In fact, ultrasound data was collected during the teaching episode at the beginning of this chapter. We can take a look at the following stills to understand how the system perceives its surroundings during the maneuver.

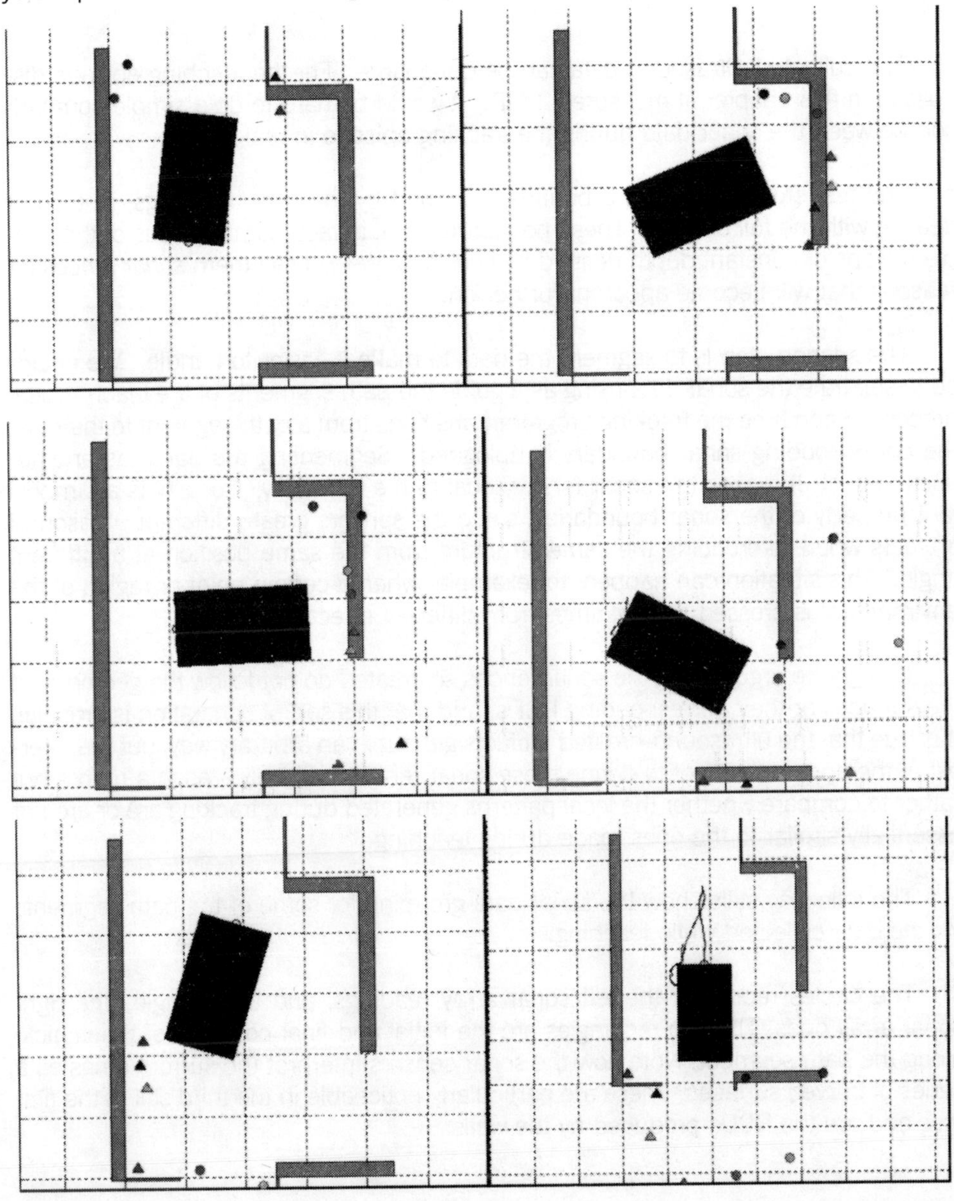

In the stills, the small triangles represent the readings of the four sensors in the right platform. The circles represent the ones in the left platform. Notice that not all sensors produce readings at all time: in the first still, for example, there are only two readings from each sensor platform. Note also how badly the ultrasound represents its surroundings. However, as we mentioned earlier, we do not care about the exactitude of the representation, just that it is roughly the same during teaching and tracking.

The collected ultrasound data can be quite large. For the teaching episode discussed in this chapter, it measures 2 MB. It would be hard to do a simple comparison between the sonar data during the tracking episode in real-time.

The first step is to locate a boundary defined by the sonar readings, instead of dealing with the full dataset. These boundaries are, in fact, combinations of the simple regions of constant depth defined by Leonard. We will call them *sonar fences* for reasons that will become apparent further on.

The second step is to segment the data to make it easier to handle. We decided to separate the sonar data using as a guide the path segments of the teach-repeat process. Each time the tracking program transitions from a path segment to the next, the corresponding sonar boundary is uploaded. Segmenting the data has another useful effect: It makes the sonar fences local to the trajectory. Locality is an important property of the sonar boundaries, since the sensors create different ultrasound artifacts when referencing the same structure from the same position at a different angle. This situation can happen, for example, when a certain point or region of the environment is crossed several times from different directions.

It might be argued that the sonar fences so created do not follow the geometry of the surfaces or their corresponding RCDs, and that this sort of separation is artificial. It is true that the ultrasound-created artifacts get cut in an arbitrary way, but the interest of the approach is not to define these sonar fences precisely – as in a map – but rather to compare whether the local patterns generated during tracking are or are not essentially similar to the ones made during teaching.

The following stills show the sonar data grouping for some of the path segments for the data collected while teaching.

The circles represent the left sonar array readings, and the triangles the right sonar array ones. The two rectangles are the initial and final positions of the vehicle along the path segment. Note how the sonar sensors interpret the surroundings as a series of curved surfaces. These are particularly noticeable in the third still of the first row, and are the RCDs produced by the walls.

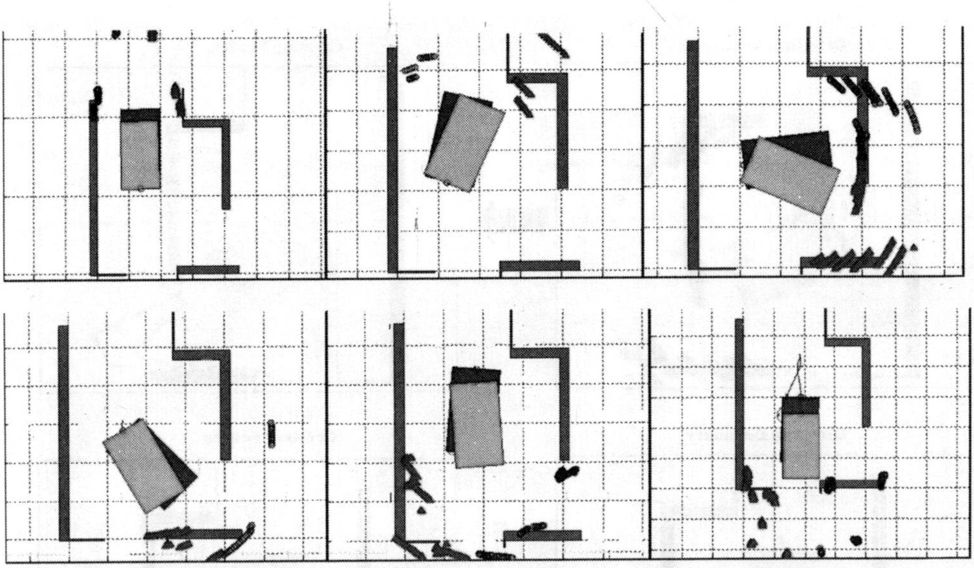

Now is possible to detect the sonar readings that are closest to the vehicle for each path segment, and use them to define a boundary. This boundary is made out of the line segments that connect the chosen sonar readings. *Figure 10-32* shows two examples of boundary detection.

In the left side of the figure, the stars are the sonar readings closest to the vehicle. In the right side, the lines are the two generated sonar fences, one for each sensor plate. These fences are defined by the extracted readings.

The process of extracting the fences for each sensor is done in the postprocessor. In that way, the burden of processing and simplifying the taught data is done offline.

Our strategy to detect obstacles needs to be simple, so that the system can perform it in real-time. At the same time it needs to be robust, so that small inevitable variations in the tracking are not perceived as changes in the environment. Before we describe our chosen approach, we can get an idea of how closely the sonar data of a tracking episode follows the taught sonar fences. The following stills correspond to ultrasound data collected during the tracking episode analyzed in this chapter, with its corresponding sonar fences superimposed.

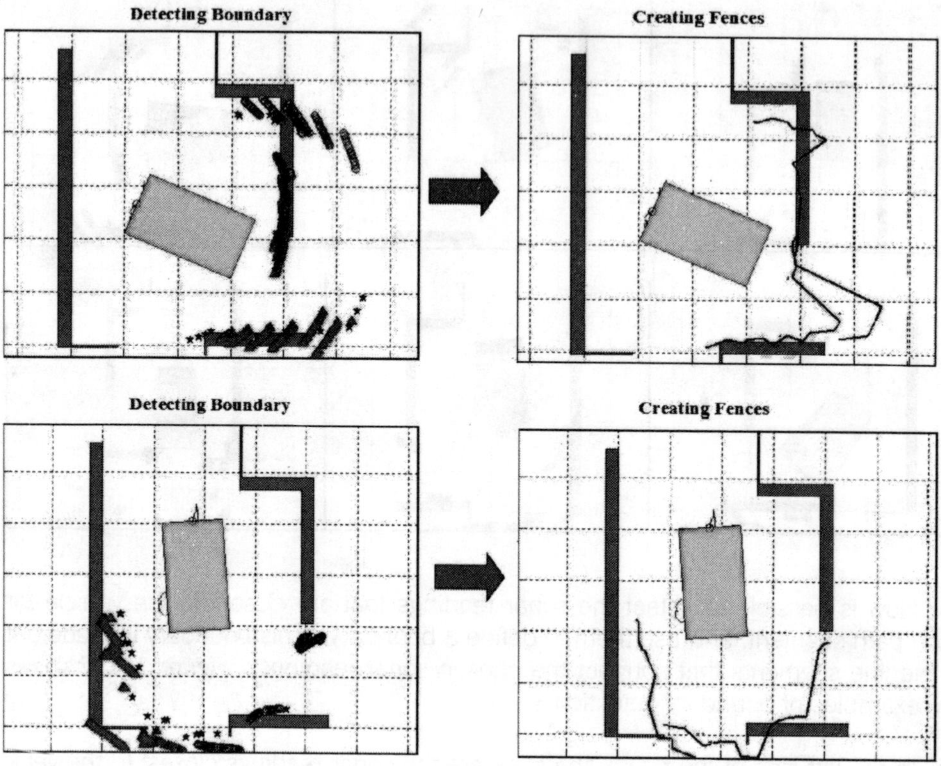

Figure 10-32. Boundary detection.

Notice how most of the sonar readings in the stills (*see opposite page*) are very close to the boundary or "outside" the fences. Informally, what we mean by "outside" is that the sonar fence is between the ultrasound reading and the wheelchair, separating them. Some readings appear "inside" the fences. This happens in slides 4 and 6. This phenomenon occurs when the trajectory tracked is not exactly the same as that which was taught. However, the sonar pattern looks roughly the same, as it should, since nothing has changed in the environment. The approach to be developed has to be robust enough to discern that these discrepancies in the pattern are not due to an obstacle.

How do the ultrasound patterns change in the presence of an object? We can get an idea by looking at *Figure 10-33*.

On the right side of the figure we have video stills of a tracking episode while a student moved in front of the wheelchair. On the left we have the plotted sonar data. Note how in both cases there is a significant distortion in the ultrasound pattern. In fact, we get readings inside the fences far from the sonar boundaries.

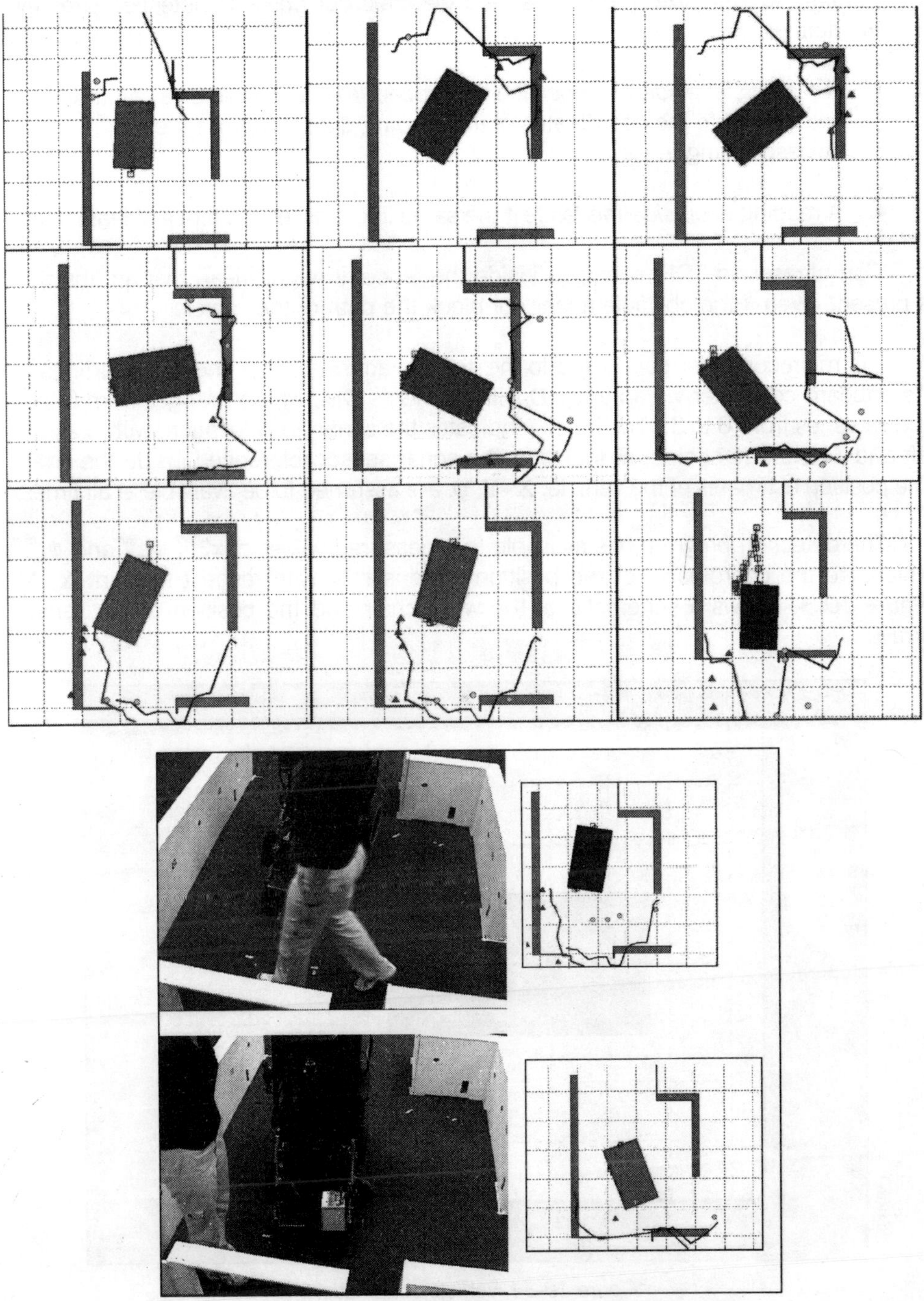

Figure 10-33. Sonar pattern in presence of an obstacle.

It is time to formalize these ideas, and describe our obstacle detection strategy. First we define:

- A reading is *inside the fence* if a line traced from the sonar array mounted on the vehicle to the coordinates of the reading does not cross the corresponding fence.

- A reading is *outside the fence* if the same line crosses the fence at least once.

If an ultrasound TOF reading is inside the fence, it might mean that an obstacle is present, even if the obstacle might not block the path of the vehicle.

The microcontroller connected to the sonar platforms sends the TOF readings to the onboard computer via an array **D** with elements D_{lm}, where l ranges from 0 to 1 and m varies from 0 to 3. The index l indicates the sonar array location, with 0 as the left and 1 as the right side; m indicates the sonar sensor referenced inside the array. The position estimates of the vehicle, X, Y, ϕ, are assumed to be available at all times.

The detection program has available the measured values of x_{lm}^{offset}, y_{lm}^{offset} and ϕ_{lm}^{offset} which are the coordinates of the position of sensor l_m with respect to point X, Y. *Figure 10-34* shows a schematic of the wheelchair and the position of the sensor arrays.

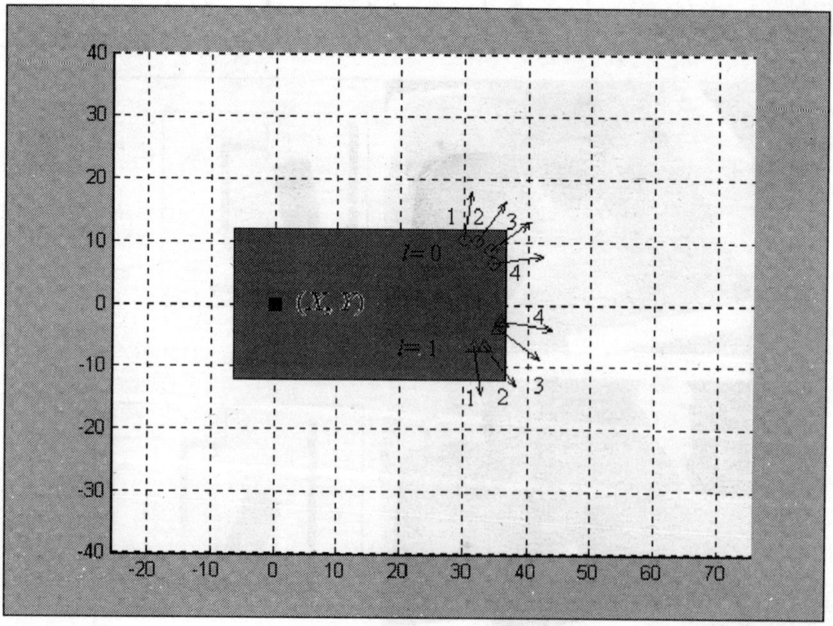

Figure 10-34. Vehicle schematic.

In the figure, the rectangle represents the vehicle and the dark square is the mid-point of the axis between the two drive wheels. The circles symbolize the sonar sensors on the left footrest and the triangles the ones on the right. The arrows show the orientation of the transducers and the direction of the sonar emission. The numbers indicate the value of m for each sensor in the array.

The values of the sensor offsets are:

$$\mathbf{x}_0^{offset} = (36.375,\ \ 38.5,\ 40.5,\ 41.0625)^T,$$
$$\mathbf{y}_0^{offset} = (10.5,\ 10.1875,\ 8.8125,\ 6.75)^T,$$
$$\phi_0^{offset} = (1.425,\ 0.9372,\ 0.6335,\ 0.1448)^T,$$

for the left footrest, and

$$\mathbf{x}_1^{offset} = (38.125,\ 39.75,\ 41.875,\ 42.3125)^T,$$
$$\mathbf{y}_1^{offset} = (-6.687,\ -6.625,\ -3.9375,\ -2.625)^T,$$
$$\phi_1^{offset} = (-81.7,\ -53.7,\ -36.3,\ -8.3)^T,$$

for the right. The angle measures are in radians and the relative position of the sensors is in inches.

First we need to calculate the position of each sensor with respect to point ***A***, the center of the axle. These coordinates are computed with respect to the absolute reference frame, and they are called xs_{lm}^0, ys_{lm}^0 at the point of emission, and xs_{lm}^1, ys_{lm}^1 at the endpoint of the sonar TOF reading (also called *point of return*). The equations to calculate these coordinates are

(10.43)
$$\begin{aligned}
xs_{lm}^0 &= x_{lm}^{offset}\cos\phi - y_{lm}^{offset}\sin\phi + X\\
ys_{lm}^0 &= x_{lm}^{offset}\sin\phi + y_{lm}^{offset}\cos\phi + Y\\
xs_{lm}^1 &= (x_{lm}^{offset} + D_{lm}\cos\phi_{lm}^{offset})\cos\phi - (y_{lm}^{offset} + D_{lm}\sin\phi_{lm}^{offset})\sin\phi + X\\
ys_{lm}^1 &= (x_{lm}^{offset} + D_{lm}\cos\phi_{lm}^{offset})\sin\phi + (y_{lm}^{offset} + D_{lm}\sin\phi_{lm}^{offset})\cos\phi + Y
\end{aligned}$$

They are the result of applying a rotation and a translation with respect to reference point $\mathbf{X}$, $\mathbf{Y}$, ϕ.

The endpoint of the sonar reading will be compared to the segments of the sonar fence, formed by consecutive points $(x_{li}^{fence}, y_{li}^{fence})$ and $(x_{li+1}^{fence}, y_{li+1}^{fence})$. Here the index i ranges from 0 to I_l-1, and I_l is the total number of points of fence l. Refer to *Figure 10-35.*

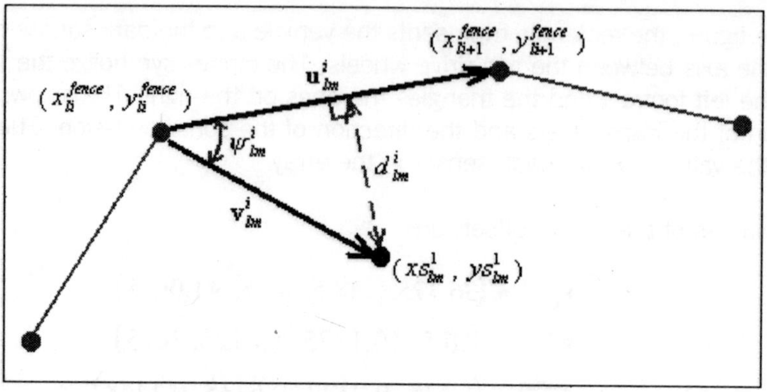

Figure 10-35. Fence Comparison.

In the figure the dark circle represents the return of the sonar reading, and the lighter circles represent points of the fence. To find the shortest distance of point $(xs_{lm}^{1}, ys_{lm}^{1})$ to the indicated fence segment, two vectors, $\mathbf{u}_{lm}^{i}$ and $\mathbf{v}_{lm}^{i}$ are defined. The displacement vector $\mathbf{v}_{lm}^{i}$ is the vector from point l_i of the segment to the endpoint of the sonar reading. The vector that joins two consecutive points of the fence $(x_{li}^{fence}, y_{li}^{fence})$ and $(x_{li+1}^{fence}, y_{li+1}^{fence})$ is denoted as $\mathbf{u}_{lm}^{i}$. Then

(10.44)
$$\begin{aligned}\mathbf{v}_{lm}^{i} &= \left(xs_{lm}^{1} - x_{li}^{fence},\ ys_{lm}^{1} - y_{li}^{fence}\right)^{T}\\ \mathbf{u}_{lm}^{i} &= \left(x_{li+1}^{fence} - x_{li}^{fence},\ y_{li+1}^{fence} - y_{li}^{fence}\right)^{T}.\end{aligned}$$

The angle between these two vectors can be calculated using the dot product. This angle is denoted as ψ_{lm}^{i} in *Figure 10-35.* By simple vector algebra,

(10.45)
$$\cos\psi_{lm}^{i} = \frac{\mathbf{v}_{lm}^{i} \cdot \mathbf{u}_{lm}^{i}}{\left|\mathbf{v}_{lm}^{i}\right|\left|\mathbf{u}_{lm}^{i}\right|}.$$

This angle is useful to determine the distance d_{lm}^{i} of the sonar reading l_m to fence element l_i.

There are three possible cases to consider, based on the relative situation of the return position in space and the fence endpoints. In the first case, the angle between $\mathbf{u}_{lm}^{i}$ and $\mathbf{v}_{lm}^{i}$ is larger than 90 degrees. That condition can be rephrased by defining three vectors r_1, r_2 and r_3 with origin in $\mathbf{X} = 0$, $\mathbf{Y} = 0$, such that

$$|r_1| = \sqrt{(xs^1_{lm})^2 + (ys^1_{lm})^2}$$

(10.46)
$$|r_2| = \sqrt{(x^{fence}_{li})^2 + (y^{fence}_{li})^2}\ .$$

$$|r_3| = \sqrt{(x^{fence}_{li+1})^2 + (y^{fence}_{li+1})^2}$$

In this setting $|r_2| < |r_3|$ by definition, since the fence points (x^{fence}_{li} , y^{fence}_{li}) and (x^{fence}_{li+1} , y^{fence}_{li+1}) are ordered from shortest to largest distance from the origin. The condition can be expressed as $|r_2|<|r_3|$. This particular situation produces a negative value of $\cos\psi^i_{lm} \bullet d^i_{lm}$ is defined as the distance between the point of sonar return *lm* and the first point of fence element *li*. Refer to *Figure 10-36*.

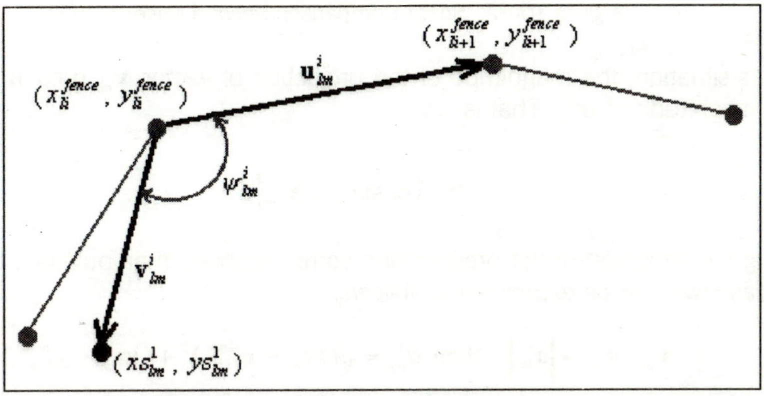

Figure 10-36. Sonar comparison first case.

Since the sign of the cosine is determined by the numerator of equation (10.45), the condition for the first case can be rewritten as,

(10.47)
$$\text{if } \mathbf{v}^i_{lm} \cdot \mathbf{u}^i_{lm} < 0, \text{ then } d^i_{lm} = |\mathbf{v}^i_{lm}|.$$

Use of the dot product instead of the cosine improves the performance of the comparison from a computational point of vue, since it executes two multiplications and an addition instead of an approximation of an infinite series.

In the second case, the sonar return *lm* has a larger magnitude than the last point of fence element *li*. That is, $|r_1| > |r_3|$ with $|r_2| < |r_3|$ as defined in equations (10.46), and d^i_{lm} is defined as the distance between the point of sonar return *lm* and the last point of fence element *li*. Refer to *Figure 10-37*.

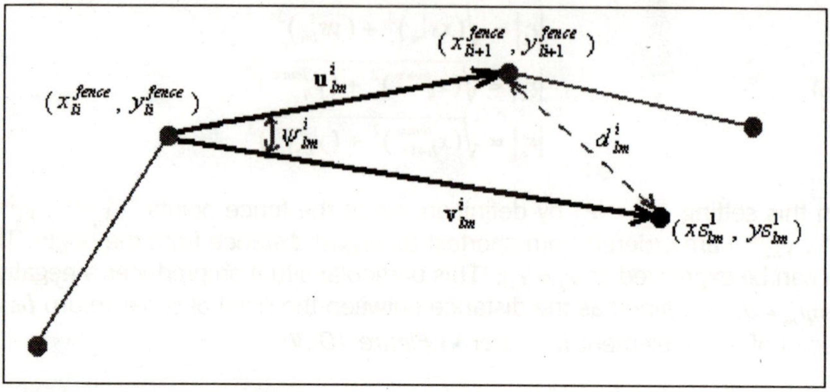

Figure 10-37. Sonar Comparison Second Case.

In this situation, the magnitude of the projection of vector $\mathbf{v}_{lm}^{i}$ onto $\mathbf{u}_{lm}^{i}$ is larger than the magnitude of $\mathbf{u}_{lm}^{i}$. That is

(10.48) $$\left|\mathbf{v}_{lm}^{i}\right| \cos\psi_{lm}^{i} > \left|\mathbf{u}_{lm}^{i}\right|.$$

Using the definition of dot product and some algebraic manipulation, the condition for case two can be expressed as follows:

(10.49) $$\text{if } \mathbf{v}_{lm}^{i} \cdot \mathbf{u}_{lm}^{i} > \left|\mathbf{u}_{lm}^{i}\right|^{2}, \text{ then } d_{lm}^{i} = \sqrt{(xs_{lm}^{1} - x_{li+1}^{fence})^{2} + (ys_{lm}^{1} - y_{li+1}^{fence})^{2}}\,.$$

Finally, in the third case the sonar return lm has a magnitude between the two points that define fence element li. That is $|r_2| < |r_1| < |r_3|$ as defined in equations (10.46), and d_{lm}^{i} is defined as the distance between the point of sonar return lm and the line segment defined by fence points $x_{li}^{fence}, y_{li}^{fence}$ and $x_{li+1}^{fence}, y_{li+1}^{fence}$. This is the case shown in *Figure 10-35.* The distance from the sonar reading to the fence segment is

(10.50) $$d_{lm}^{i} = \left|\mathbf{v}_{lm}^{i}\right| \sin\psi_{lm}^{i}\,.$$

By the definition of cross product and some algebraic manipulation, the distance can be calculated as

(10.51) $$d_{lm}^{i} = \frac{\left|\mathbf{u}_{lm}^{i} \times \mathbf{v}_{lm}^{i}\right|}{\left|\mathbf{u}_{lm}^{i}\right|}.$$

This distance calculation has to be done for each segment of the fence. In the end, the measure of the distance of the sonar return lm to fence l is

(10.52) $$d_{lm}^{fence} = \min(d_{lm}^{1}, d_{lm}^{2}, \ldots, d_{lm}^{I_l-1}).$$

Now that we know the distance to the fence, we need to find on which side of the fence is the detected sonar reading. In order to perform that test, the sonar segment lm defined by points $(xs_{lm}^{0}, ys_{lm}^{0})$ and $(xs_{lm}^{1}, ys_{lm}^{1})$ is extended until it intersects the extended fence segment li, defined by points $(x_{li}^{fence}, y_{li}^{fence})$ and $(x_{li+1}^{fence}, y_{li+1}^{fence})$ at point $(x_{li}^{inter}, y_{li}^{inter})$. Refer to *Figure 10-38*.

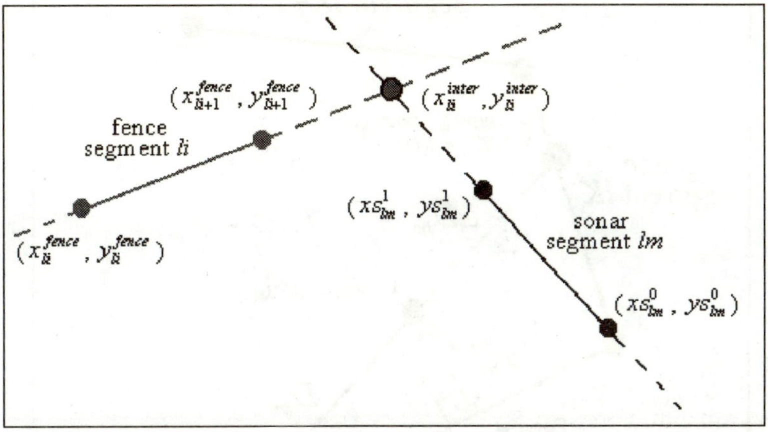

Figure 10-38. Intersection of extended sonar reading and fence.

The solution for this system of equations for the point of intersection of two lines is

(10.53)
$$x_{li}^{inter} = \frac{m_{lm}\, xs_{lm}^{1} - m_{li}^{fence}\, x_{li+1}^{fence} - ys_{lm}^{1} + y_{li+1}^{fence}}{m_{lm} - m_{li}^{fence}}$$
$$y_{li}^{inter} = \frac{m_{li}^{fence}\left[m_{lm}\left(x_{li+1}^{fence} - xs_{lm}^{1}\right) - ys_{lm}^{1}\right] - m_{lm}\, y_{li+1}^{fence}}{m_{li}^{fence} - m_{lm}}$$

where the slopes of the extended line segments are

(10.54)
$$m_{lm} = \frac{ys_{lm}^{0} - ys_{lm}^{1}}{xs_{lm}^{0} - xs_{lm}^{1}}$$
$$m_{li}^{fence} = \frac{y_{li+1}^{fence} - y_{li}^{fence}}{x_{li+1}^{fence} - x_{li}^{fence}}.$$

Assume that for some $i = K$ the point $(x_{lK}^{inter}, y_{lK}^{inter})$ is on fence segment lK. Then the following conditions would be met:

(10.55)
$$\min(x_{lK}^{fence}, x_{lK+1}^{fence}) \le x_{lK}^{inter} \le \max(x_{lK}^{fence}, x_{lK+1}^{fence})$$
$$\min(y_{lK}^{fence}, y_{lK+1}^{fence}) \le y_{lK}^{inter} \le \max(y_{lK}^{fence}, y_{lK+1}^{fence}).$$

This test is performed independently of the distance calculation, since the segment where d_{lm}^{i} is minimal might not be the same as the one where condition (10.55) is satisfied. Refer to *Figure 10-39.*

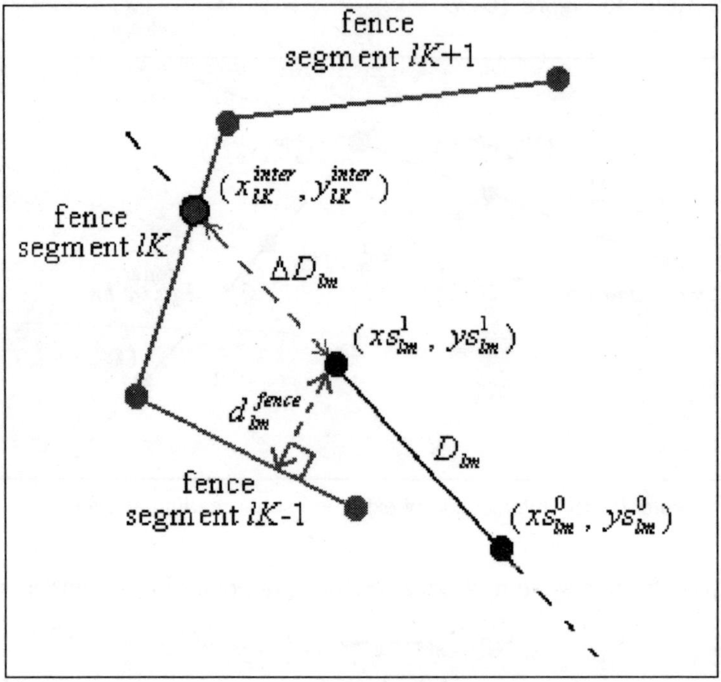

Figure 10-39. Intersection of prolonged sonar reading with fence segment.

In the figure, note how $d_{lm}^{fence} = d_{lm}^{K-1}$ but the comparison to determine on which side of the fence the sonar reading falls is performed with segment ***lK***. Note that the 'side' test takes into account the point of sonar emission and the distance calculation does not. For segment ***lK*** the distance from the sonar emission point to the point of intersection with the fence is

(10.56)
$$d_{lK}^{inter} = \sqrt{(xs_{lm}^{0} - x_{lK}^{inter})^2 + (ys_{lm}^{0} - y_{lK}^{inter})^2}\ ;$$

and the difference between the sonar TOF reading and this calculated distance is

(10.57)
$$\Delta D_{lm} = D_{lm} - d_{lK}^{inter}$$

If $\Delta D_{lm} < \mathbf{0}$, the sonar reading is inside the fence. When this happens, the program checks whether

(10.58) $$d_{lm}^{fence} > \text{DIST\_BLOCK\_MAX}.$$

If that condition is met, the disturbance might mean that an object is distorting the sonar pattern.

For the purpose of the *CPWNS*, each time the program detects condition (10.58), it adds +1 to an internal counter. When a sonar reading meets this condition, it is called a *blocked reading*. There is a counter that keeps track of the left side and one of the right. If any of the counters reaches the value SAMPLES_STOP counts, the program assumes that there is an obstacle. At that stage, the system is in a *locked* state.

Note that the possible obstacle might not be blocking the trajectory, since this scheme only detects whether the sonar pattern generated during the tracking episode displays no major deviations from the fences calculated during the teaching phase.

Once the system is in a locked state, each unblocked reading will subtract a count from its corresponding side, while adding for each blocked reading. If the counter reaches the value of SAMPLES_RUN, the system *unlocks*. The system will be in a locked state if either of the counters has a value of SAMPLES_STOP, and will unlock only if both counters have a value of SAMPLES_RUN. The counters are bounded by these two constants and they start with an initialization value of SAMPLES_RESET.

The current setting of these constants is SAMPLES_RUN = -3, SAMPLES_RESET = 0 and SAMPLES_STOP = 2. These can be changed depending on the desired sensibility of the system, as long as

(10.59) SAMPLES_RUN < SAMPLES_RESET < SAMPLES_STOP.

What remains to be determined is the tolerance parameter DIST_BLOCK_MAX. This proved to be quite difficult, since strong reflections produced by an edge, combined with a slight deviation in the chair's angular position, can produce strong readings inside the fence. Observe *Figure 10-40*.

In the figure, the edge in the lower right part produces noticeable readings inside the fence with no obstacle present. In most other geometric configurations, the sonar readings stay well outside the fence. Since sharp edges are very common in real home environments, a solution is needed.

A possible approach is to define two values of DIST_BLOCK_MAX. In the case of segments with no sharp edge reflection we calibrated DIST_BLOCK_MAX to be 6

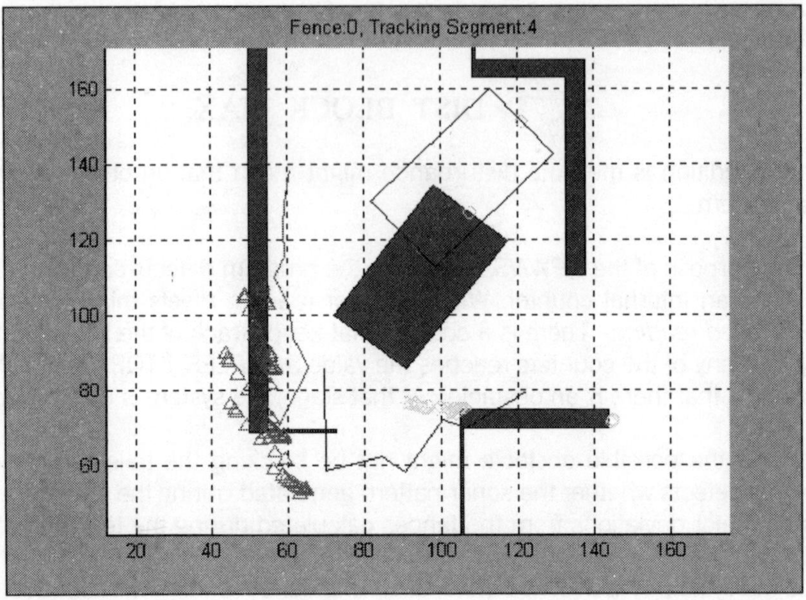

Figure 10-40. Strong corner reflection.

inches, to account for small variations in the path. This value permits enough sensitivity to detect a book lying flat on the floor, represented by a small rectangle in *Figure 10-41*.

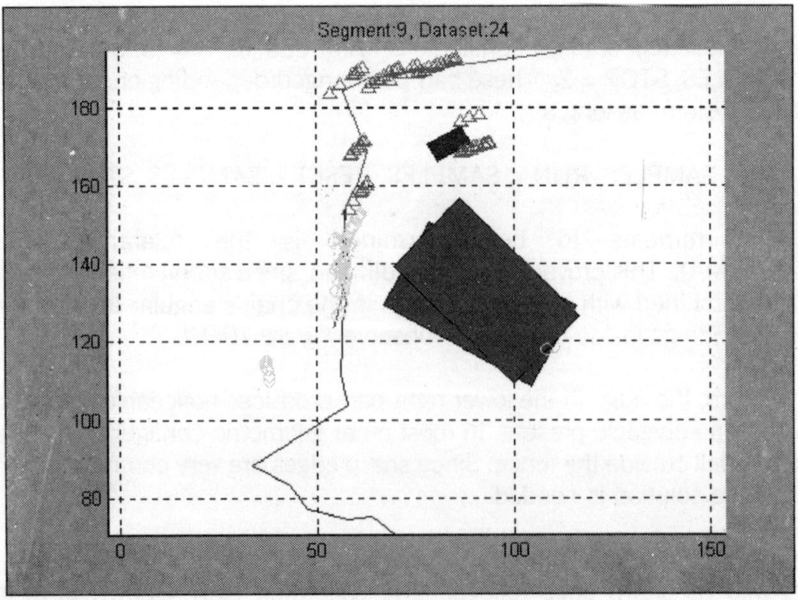

Figure 10-41. Sonar detection of a book.

It is also enough to detect the legs of a stool, something hard to achieve since the reflective surface is rather small. In *Figure 10-42*, the stool is represented by a square.

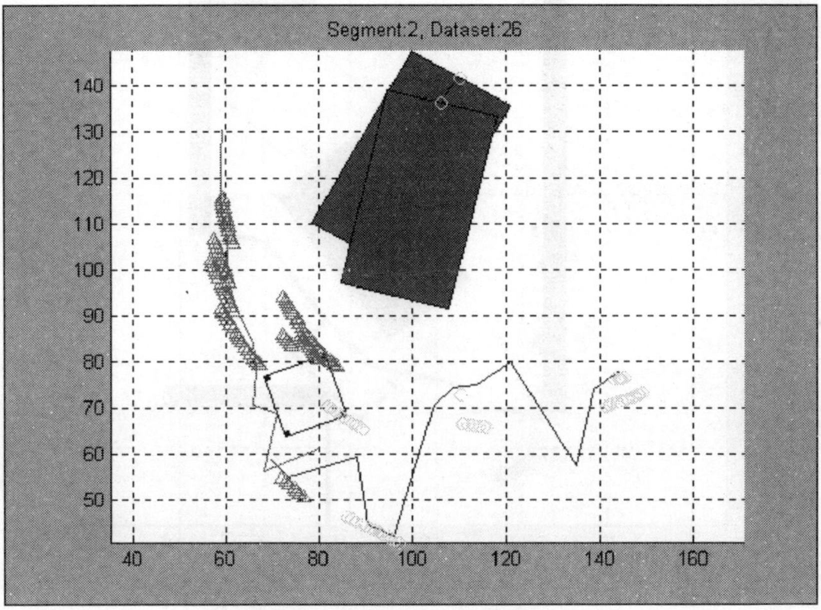

Figure 10-42. Sonar detection of a stool.

In the case of segments where the wheelchair detects strong reflections, like the one shown in *Figure 10-40*, the value DIST_BLOCK_MAX can be as high as 12 inches. These values were obtained empirically, by having the system run through different paths with and without obstacles. At this stage, we set these values manually. By manually we mean that we modify DIST_BLOCK_MAX for a certain path segment if the system locks with no obstacle present during the first trial tracking. The process could be automated in the postprocessor, since these strong reflections can be detected in the teaching data.

Another practical solution that we have been exploring lately is to move the sonar boundaries in the path segments that might include sharp edges. How do we achieve that behaviort? We simply put a board or a plank in front of a sharp edge. This creates a *virtual* fence that is closer to the vehicle. The undesired strong reflection can be made to stay always outside the fence, by generating a tighter virtual fence. Observe *Figure 10-43.*

The figure shows a similar sonar profile as the one in *Figure 10-40*. By putting a box in the problematic corner during the teaching maneuver we have created a tighter sonar boundary. This box is symbolized by the dashed rectangle in the figure. The box is removed after the teaching step, but the new boundary remains for the subse-

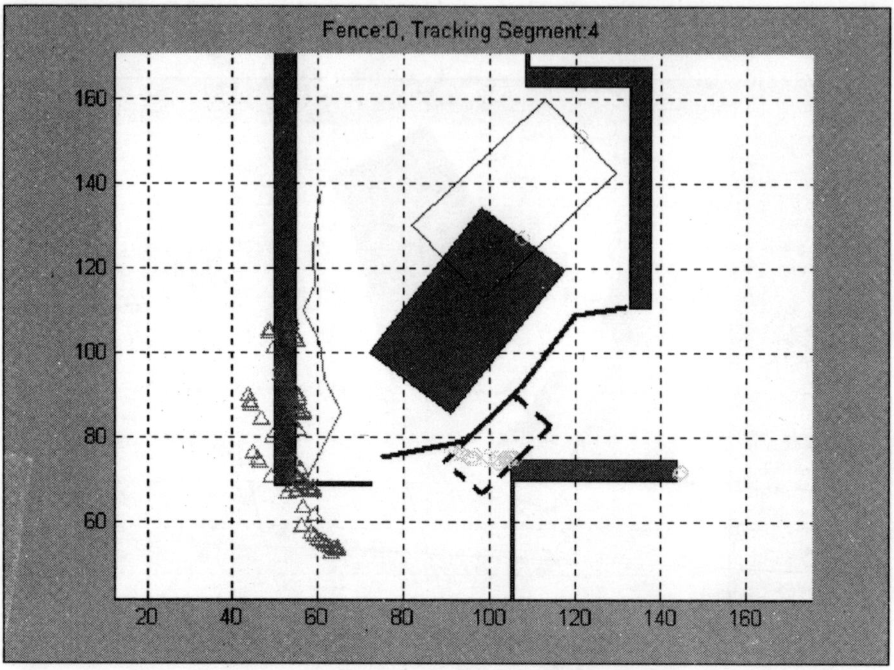

Figure 10-43. Tracking with a Tighter Fence.

quent tracking episodes. The system can still detect an obstacle while tracking, as long as it is inside the new boundary. Using the idea of a virtual fence, we can limit as much as we want the range distance where an obstacle would be detected. During the experiments at Hines, VA Hospital, we have used boxes and wood planks to create tighter virtual fences.

We have discussed how the system enters a locked state after detecting an obstacle, but not what action it takes. This action is actually quite simple: when the system locks, the vehicle stops. When it unlocks, the vehicle resumes its tracking along the path.

This naïve approach can be quite powerful, and it becomes useful when combined with the ability to reverse a trajectory. Once an obstacle is detected, the system could ask the user whether they want to go back to the starting point of the path. This action has still not been implemented in the *CPWNS*, and the re-tracing of a path with rotational and linear segments needs further testing (although reverse tracking of a taught path in forward or backward directions has been tested extensively and works well).

The obstacle detection capability is the groundwork to perfect and develop an eventual strategy for obstacle avoidance. There are several existing systems that have strategies of obstacle avoidance. Some of them can even detect moving objects, and plan their trajectories accordingly. The difference in terms of the parameters of the problem with the *CPWNS* is that these others systems normally work in big rooms with lots of space to maneuver. For example, the *MAid* prototype was tested in the central railway station of Ulm and in the exhibition halls of the Hannover Messe '98, the largest industrial world fair. In contrast, the *CPWNS* can move in very cramped environments with little space to maneuver.

Consider again the trajectory discussed throughout this chapter. If an object was placed at the center of the environment, there would be no alternative route for the vehicle to complete the trajectory, since the walls are simply too close. In a case like that, stopping and retracing the path might be the only option. Maybe in later versions the system could compare both approaches, and if the space is large enough, it could try to circumnavigate the blocking object.

In any case, the sonar strategy described does an excellent job detecting obstacles. What is interesting about the scheme is that it does not need a preprogrammed map. Also, it does not need a complicated algorithm to categorize complex RCDs, and the creation of the fences is done offline. Using the simple idea that the sonar patterns are consistent, it achieves a competent level of robustness. This phenomenon is also a direct consequence of the coarseness of the ultrasound patterns. It is unlikely that this scheme would work if sonar behaved like a ray-trace scanner.

More importantly, this sonar scheme only works because it is used in combination with teach-repeat. Without the teaching step, we would need to model the environment in some way, and we have seen how extremely difficult that task can be.

Thoroughout this chapter we have seen the *CPWNS* performing in a laboratory environment. Yet, we have mentioned several times the importance of taking these applications outside the shelter of a university setting and into the "real" world. For that purpose, we have set two environments at the Hines, VA Hospital. One is a test environment provided by the hospital that mimics a real apartment. This setting was constructed in order to help patients who have recently suffered a crippling spinal cord injury. Our only addition was the cues set in a bedroom, a long hallway, a living room and a kitchen.

The second test environment comprises a large room, a hallway, and an office. Again, our only addition was the cues spread throughout. The following stills show the wheelchair tracking a path in the hospital environment.

In the study done at the Hines VA Hospital, six disabled individuals were taught how to use the *CPWNS*. All subjects were able to master the system capabilities for moving between destinations in 15 to 20 minutes. In parallel, two undergraduate stu-

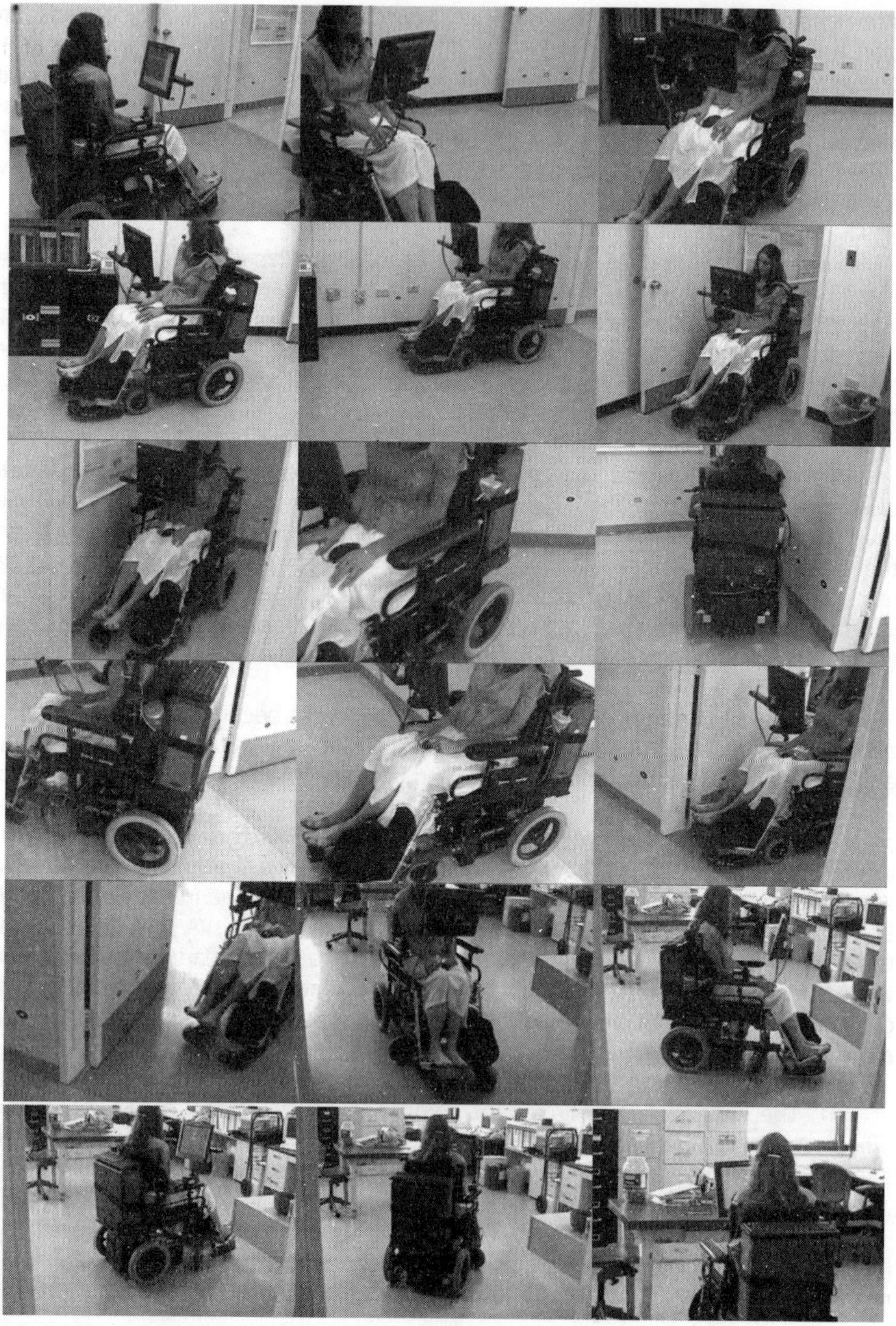

dents at the University of Notre Dame, Timothy J. Sheehan and Kristen A. Woyach, were trained at the Dexterity, Vision and Control Laboratory. They had to learn how to achieve the tasks of teaching, capturing the sonar templates, setting up and environment, path-generation and general testing. In both cases, within three one-hour sessions, the students were confident enough to use the full capability of the system unsupervised.

In addition, we have taken the *CPWNS* on the road, that is, we have showed the system at different venues and presentations. For example, we did live demonstrations at the Indiana State Fair 2002 (as part of the Indiana 2016 booth), at the RESNA 26th International Conference and at the Supercomputing 2003 Conference (at the *Research in Indiana* booth). We would bring the wall segments shown in *Figures 10-20* and *10-21* to reuse the paths previously taught in the laboratory. Some of these presentations required for the *CPWNS* to run non-stop for 5 hours or more. Of course, we could not predict the lighting conditions, the texture of the floor or the weight of people who decided to test-drive the system. The system performed admirably under such stress.

These experiences give us hope that extending teach-repeat to nonholonomic systems is indeed a robust, viable approach for mobile robotics. We believe it can be applied to real-world problems, outside of the shelter of the university. The application of the wheelchair described herein, has the potential to significantly improve the life of severely disabled individuals. As has been pointed out at the beginning of this chapter, this type of patients has currently no alternative means of locomotion.

However, to just concentrate in this particular application would be missing the forest for the trees. The approach discussed at length in this chapter has the potential to finally fulfill the promise of autonomous navigation. A robot of this type, could deliver mail in a large office building, move supplies to a desired site, clean and maintain floors in airports or factories, even dock boats in a pier. All these applications are possible with current technology!

Of course, teach-repeat has limitations. A mobile robot of this type is dependent on a human to teach it the trajectory to follow. As a consequence, this paradigm could never be used to explore new terrain or to control a robotic probe on another planet, for example. However, we propose another idea that could be used for that purpose in Chapter 11.

We believe that a *robotics revolution*, envisioned by the visionaries of the early twentieth century, re-imagined later by the utopists of the fifties and finally almost laid to rest by the disappointments of the late eighties and nineties could still happen. Extending teach-repeat to nonholonomic robots could finally provide us with practical, autonomous navigational systems for everyday or industrial applications.

CHAPTER 11

MOBILE CAMERA-SPACE MANIPULATION;
REALIZING MANEUVER OBJECTIVES IN THE REFERENCE FRAMES OF MOVING CAMERAS

Placement of a holonomic arm onto a wheeled base expands indefinitely the workspace of that arm. If cameras are transported with the base CSM can in principle be brought to bear on maneuver control. The question becomes: "Can we exploit the nonholonomic degrees of freedom of the wheels to lower the required dexterity - or number of effective degrees of freedom - of the onboard arm?"

Consider the forklift-type autonomous robot of *Figure 11-1*. The completely autonomous task of this wheeled robot is to engage precisely each of three boxes using a two-prong device inserted into two narrow slots beneath each box. The system then lifts the box and drives it to where visual cues like those on the engaged box have been fixed onto an empty table, as shown in the figure. The robot autonomously places the first box in alignment with those cues, and returns twice to the first table to do the same with each of the two remaining boxes. The result is a neat stack of three boxes on the second table as indicated in the final picture of *Figure 11-1*. Video of this entire sequence can be viewed at **http://www.nd.edu/~sskaar/**.

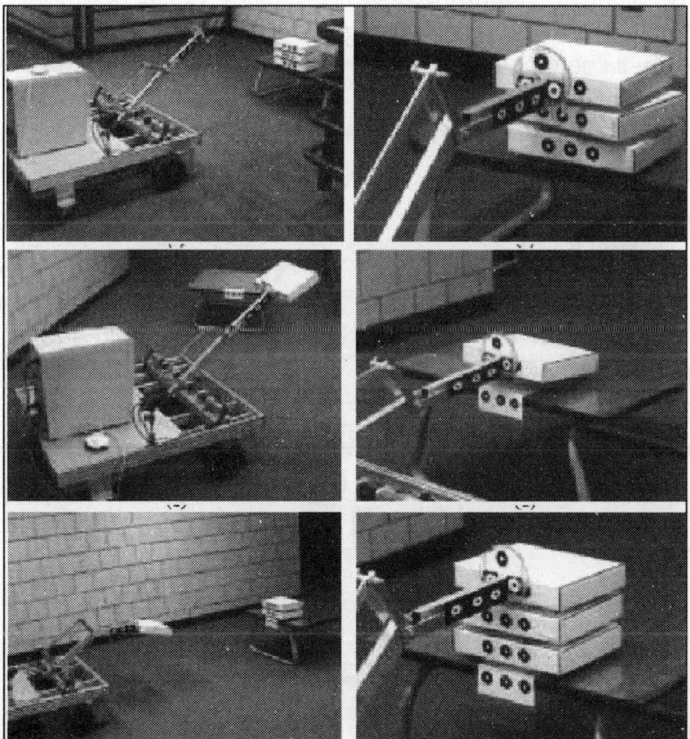

Figure 11-1. Two onboard holonomic degrees of freedom combine with two nonholonomic degrees of freedom to complete a four-degree-of-freedom task, using maneuver control in the reference frames of two onboard cameras.

Of particular interest here is the number of degrees of freedom of the task itself. Because the desired positioning never requires the boxes to depart from an orientation that is parallel to the floor, that number is four. Three components of each box's position must be controlled and one component of orientation. It's apparent from *Figure 11-1* that this particular robot has at least one more degree of freedom effectively available because an engaged box is shown being transported at an orientation clearly not parallel to the floor.

How is it, then, that, as indicated in the caption of *Figure 11-1*, a system with four servomechanisms – four degrees of freedom – can produce five degree-of-freedom results? The answer lies in the nonholonomic nature of the system's kinematics. *Figure 11-2* presents two bird's-eye-view, closely spaced poses of the base as it approaches one of the boxes. We call the forward rotation of the left wheel θ_1, as indicated in *Figure 11-3*, and the forward-rotation angle of the right wheel θ_2. Between the two poses of the platform of *Figure 11-2* the rotational increment of the right wheel, $\Delta\theta_2$, is small, positive and slightly greater than the rotational increment, $\Delta\theta_1$, of the left wheel. It is apparent from this figure that the middle cue for instance, cue **B**, will occupy slightly different in-plane coordinates x_B, y_B than it did before these incremental wheel rotations. This change is not because cue **B** moved but rather because the x-y-z coordinate system of the moving base, shown in *Figures 11-2* and *11-3,* shifts with wheel rotation.

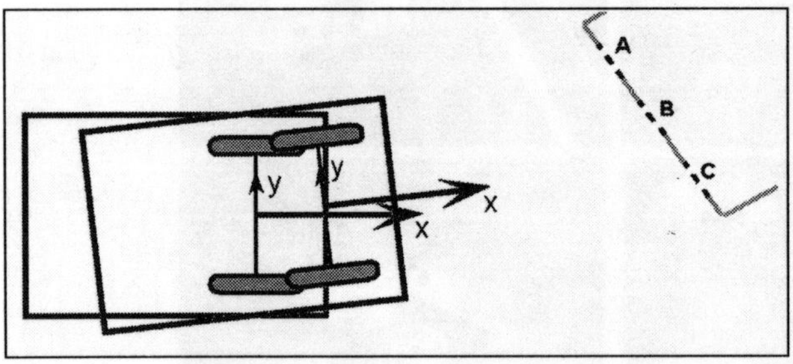

Figure 11-2. Incremental change in the relative position of the three cues with respect to the base relates to increments in wheel rotation.

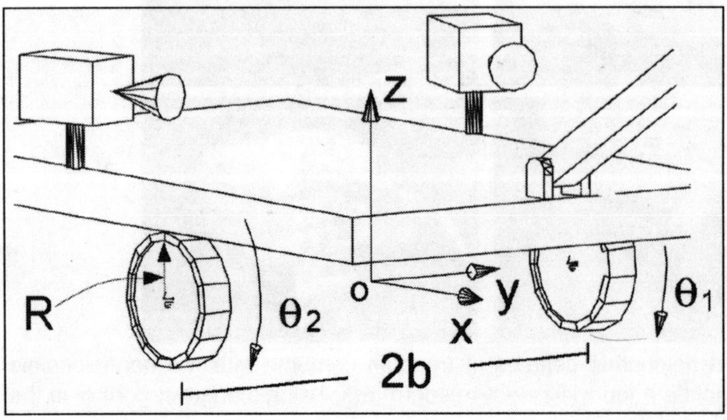

Figure 11-3. The two wheels' rotational angles are defined. Note the cameras.

The exact differential-kinematics expressions to describe the relationship between Δx_B, Δy_B, and $\Delta\theta_1$, $\Delta\theta_2$ assume no wheel slip and infinitesimal magnitudes. With these assumptions the nonholonomic kinematic expressions of interest become

$$\begin{aligned} dx_B/d\alpha &= -R + y_B R u(\alpha)/b \\ dy_B/d\alpha &= -x_B R u(\alpha)/b \\ dz_B/d\alpha &= 0 \\ d\phi/d\alpha &= -Ru(\alpha)/b \end{aligned} \tag{11.1}$$

where the introduced independent variable α and control $\mathbf{u}$ are defined as in Chapter 10 for the wheelchair:

$$\begin{aligned} d\alpha &= (\Delta\theta_1+\Delta\theta_2)/2 \\ u &= (\Delta\theta_2-\Delta\theta_1)/(\Delta\theta_1+\Delta\theta_2) \end{aligned} \tag{11.2}$$

Again, these are defined in the limit as the increments $\Delta\theta_1$ and $\Delta\theta_2$ approach zero. What is interesting, and explains the ability of just four servomechanisms to control five elements of rigid-body position/orientation, is that, taken together, Equations 11.1 and 11.2 permit the specification of wheel rotation histories, $\theta_1(t)$ and $\theta_2(t)$, that would transition for example the vehicle posed initially according to *Figure 11-4* to identically locate the fork as needed with respect to the box as indicated in *Figure 11-1*. Both in-plane position x_B, y_B as well as angle ϕ — three components in all — can be guided to three given values through judicious rotation of just two servomechanisms.

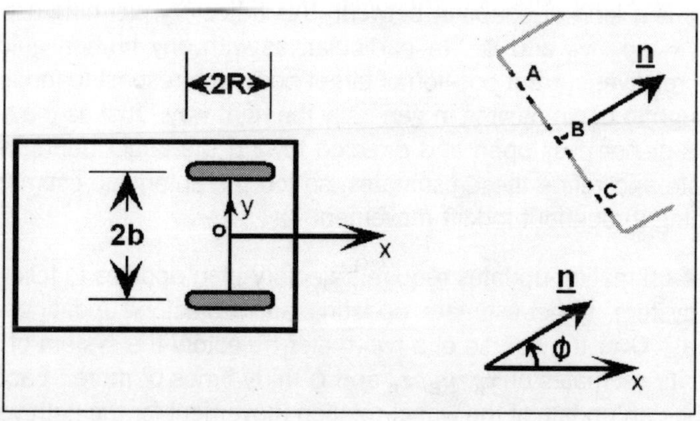

Figure 11-4. With increments in the rotations of the two wheels the coordinates of **B** with respect to the base change, as does the angle ϕ that orients the normal **n** with respect to the vehicle's x axis.

Such ability does not reside with holonomic mechanisms. But in robotics the nonholonomic advantage carries a price: As discussed in Chapter 10, it removes the property of repeatability in the sense that a simple return of the internal angles to previous rotations would guarantee return of the mechanism and delivery of a load to a corresponding earlier pose. Furthermore the "path dependence" of the relationship between wheel rotation and pose means that trajectory computation is more difficult. It isn't enough to merely designate angles θ_1 and θ_2 near the terminus; the entire *history* of wheel rotation matters to the maneuver outcome. And Mobile Camera-Space Manipulation (MCSM) was developed to address this issue. Yoder Software, Inc. (YSI) developed MCSM with the help of supprot from NASA SBIR awards, and YSI owns a patent on the technology.

Movement of the base requires answers to two questions - both of which have counterparts in the wheelchair problem of Chapter 10. The first is trajectory planning for movement of $\Delta\theta_1$ and $\Delta\theta_2$ mentioned above. And the second is estimation: what are the current values of x_B, y_B and ϕ?

Trajectory planning for the wheelchair-navigation problem of Chapter 10, involving as it does movement relative to fixed walls and an indoor environment, was entirely accomplished by a human teacher and intended for simple repeat action. In the present problem, however, we presume variability of target position/orientation comprised of location of point $\mathbf{B}$ (x_B, y_B) and of base orientation relative to the box (ϕ). So a new trajectory will, in general, need to be planned in each new situation. Not surprisingly an infinite number of trajectories will connect the current pose to the target-body-dependent terminal pose. We just need to calculate and begin to execute one of them.

There is a kind of coupling between this trajectory planning aspect and the estimation of x_B, y_B, z_B and ϕ. In particular, as with any human guidance, early notions of the relative, current position of target body with respect to forklift are rough, but good enough to begin moving in generally the right way. Just as the human operator keeps his or her eyes open and directed toward the target during approach, in order to update and refine these estimates, so too the automatic estimation process remains ongoing throughout forklift movement.

But target-estimation updates require trajectory-plan updates to follow them. For the artificial system, target-estimate updating and trajectory updating are discrete events in time. Over the course of a two-meter trajectory the system of *Figure 11-1* might update its estimates of x_B, y_B, z_B and ϕ thirty times or more. Each of these is used to produce an update of the wheel-rotation movement for the path yet to go. The latter entails defining and solving a "two-point-boundary-value problem" for $\mathbf{u}(\alpha)$ of Eqs. 11.1 and 11.2.

Many required "boundary conditions" for this determination of $\mathbf{u}(\alpha)$ are discussed below, but we first note a requirement that is associated with the desirability of nonstop, continuous and smooth motion across an estimation/path-plan transition point. Because trajectory updates need a finite amount of computer-execution time it is necessary to calculate the updates in advance — for a juncture "α" in the path that is in the near future as far as the path plan currently being executed is concerned.

Any calculated $\mathbf{u}(\alpha)$ plan under current execution has been transformed to a sequence or list of closely spaced θ_1/θ_2 rotations. These are under execution in a synchronized, nonstop way by a motion-control/motor system — in the case of the *Figure 11-1* robot, one that is commercially available. This series of θ_1/θ_2 "reference angles" (see position servomechanism discussion of Chapter 2) is computed from the $\mathbf{u}(\alpha)$ plan as follows: Beginning with a current θ_1, θ_2 and α, a small $\Delta\alpha$ is chosen and substituted for the infinitesimal $\mathbf{d}\alpha$ in the first of Eqs. 11.2. The corresponding $\mathbf{u}$ is then used from the most recent trajectory plan and current value of α, and increments for the next wheel-reference angles, $\Delta\theta_1$ and $\Delta\theta_2$, are determined through simultaneous solution of Eqs. 11.2. The new values of θ_1 and θ_2 to be added to the list are found from adding $\Delta\theta_1$ and $\Delta\theta_2$ to the previous values. And of course α is updated by adding the preselected $\Delta\alpha$ to the previous α. Because the independent variable α is proportional to distance travel by point $\mathbf{O}$ of *Figures 11-2* and *11-3*, advancing the reference path with equal intervals $\Delta\alpha$ results, for the motion-control system used, in constant speed of vehicle approach.

Fortunately, the list of θ_1/θ_2 reference values is generated much faster than the wheel-motion-control system can execute it, even though the central processor is also busy updating estimates of x_B, y_B, z_B and ϕ. The list of reference values is generated fast enough in fact that a new trajectory plan $\mathbf{u}(\alpha)$ can be determined for most of the distance left to go, a trajectory plan that will "take over" from the current one at a prescribed near-future value of α.

Let this transition value be called α_t. The trajectory plan $\mathbf{u}(\alpha)$ currently being executed has a previously calculated value of $\mathbf{u}$ at this juncture α_t. Call this $\mathbf{u}_t$. To allow for finite motor actuation to handle the transition, Newton's laws require continuity of $\mathbf{u}$ at α_t. Hence one boundary condition used in updating the trajectory-to-go is what now, for the newly calculated, future reference path, we call an "initial condition":

$$\mathbf{u}(\alpha_t) = \mathbf{u}_t \tag{11.3}$$

Note that $\mathbf{u}$ is proportional to the curvature of Point $\mathbf{O}$ in *Figures 11-2* and *11-3*. Letting α_f be the terminal value of the independent variable it is also desirable to produce straight-line motion just prior to fork insertion. Hence we impose the "final condition":

$$\mathbf{u}(\alpha_f) = 0 \tag{11.4}$$

The remaining *initial* conditions are formed using projected estimates of x_B, y_B and ϕ at what is still a future trajectory transition point, α_t. As described below the same Extended Kalman Filter algorithm used for the wheelchair problem applies camera samples of cues **A**, **B**, and **C** to update estimates of x_B, y_B, z_B and ϕ at the last-sample value of α. With numerical integration of the current trajectory plan's $\mathbf{u}(\alpha)$, we have no trouble advancing these via numerical integration of Eqs. 11.2 to α_t, the start of the newly computed trajectory plan. Calling these estimates x_{Bt}, y_{Bt}, z_{Bt} and ϕ_t, the following three initial conditions are added to the list:

$$x_B(\alpha_t) = x_{Bt}$$

(11.5) $$y_B(\alpha_t) = y_{Bt}$$

$$\phi(\alpha_t) = \phi_t$$

Three more final conditions are also added based upon the diagram of *Figure 11-4*:

$$x_B(\alpha_f) = \eta$$

(11.6) $$y_B(\alpha_f) = 0$$

$$\phi(\alpha_f) = 0$$

where, in accordance with *Figure 11-5*, η is the distance from point **0** at trajectory termination to target point B just prior to fork insertion, projected into the horizontal plane. The actual value of η is determined from a combination of the holonomic kinematics model of the onboard arm combined with the estimate of elevation z_{Bt}. Because, according to the nonholonomic kinematics model, z_B will not change from α_t to α_f, the two on-board angles θ_3 and θ_4 are evaluated such that the fork is parallel to the floor and the elevation is two cm above the estimate of $z_B(\alpha_f)$. After arriving at α_f the vehicle advances one box depth and lowers the box to exactly $z_B(\alpha_f)$ prior to backing out the fork.

Two issues remain. First, it seems as if the nominal forward kinematics have been used freely throughout the above development despite the assertions elsewhere in this book that they are not to be trusted, even for holonomic robots, especially where computer vision is involved in determining target points. And second, we have not indicated how the actual reference path $\mathbf{u}(\alpha)$, $\alpha_t<\alpha<\alpha_f$, can be generated such that it satisfies all of Eqs, 11.1-11.6. We consider this second matter first.

There are many ways to approach the problem of generating a trajectory plan that satisfies the state equations, Eqs. 11.1-11.2, together with the boundary conditions of Eqs. 11.3-11.6. One approach is to choose a "performance index" and find a path or trajectory that is optimal in the sense that it minimizes that index. Another is to force the trajectory to conform to some kind of functional form such as a polynomial form and choose coefficients of that form so as to satisfy all of the conditions.

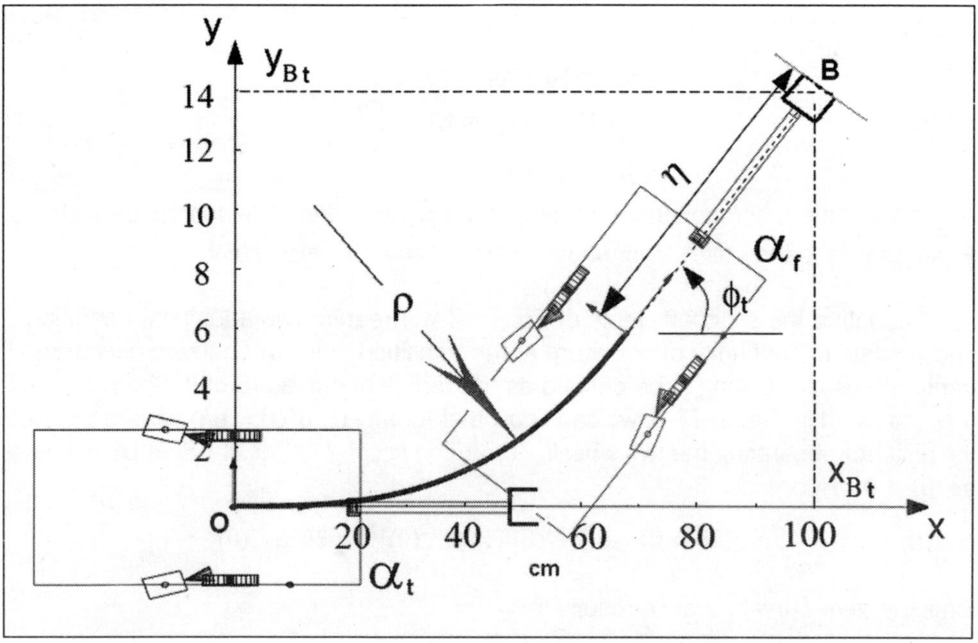

Figure 11-5. Bird's eye view of the vehicle at boundary points α_t and α_f, with definition of η and ρ.

Perhaps the simplest example of the latter can best be illustrated by referring to *Figure 11-5* above. Rather than working with the position of stationary point **B** with respect to the moving vehicle frame, consider the position of the moving base's point **0** with respect to the base's own frame at α_t. Call this $\mathbf{x_0}$, $\mathbf{y_0}$. Let the trajectory plan be specified in terms of $\mathbf{y_0(x_0)}$ according to the following polynomial:

(11.7) $$y_O(x_O) = a_0 + a_1 x_O + a_2 x_O^2 + a_3 x_O^3 + a_4 x_O^4 + a_5 x_O^5$$

From the initial pose to the left in *Figure 11-5*, the following pair of initial conditions can be surmised:

(11.8) $$y_O(0) = 0$$ $$y_O'(0) = 0$$

The second of these is actually a form of the nonholonomic kinematic constraint of the wheeled vehicle (see Chapter 9).

Referring to the right pose in *Figure 11-5*, we have:

(11.9)
$$x_{Of} = x_{Bt} - \eta\cos(\phi_t)$$
$$y_{Of} = y_{Bt} - \eta\sin(\phi_t)$$
$$y_{Of}' = \tan(\phi_t)$$

where $\mathbf{x_{of}}$ is the $\mathbf{x}$-component of the terminal position of the base coordinate system's origin relative to the base's coordinate system at the transition point.

Evaluating the six coefficients of Eq. 11.7 will require two additional conditions. These relate to continuity of curvature at the transition point α_t and zero curvature at the terminus α_f. Letting ρ be defined as the radius of curvature of the path of point $\mathbf{0}$ as indicated in *Figure 11-5*, we can show that for any α: $\mathbf{u}(\alpha) = \mathbf{b}/\rho$, where $\mathbf{b}$ is half the distance separating the two wheels. From the second of Eqs. 11.8, at α_t, we have $\mathbf{yo'(0) = 0}$. Hence:

(11.10)
$$u_t = 2b/\rho(0) = b\ y_O''(0)/[1+y_O'^2(0)]^{3/2} = 2b\ y_O''(0)$$

Similarly, zero curvature at α_f results in:

(11.11)
$$y_O''(x_{Of}) = 0$$

Substitution of the boundary conditions in the form of Eqs. 11.8-11.11 into the polynomial of Eq. 11.7 results in six equations for six unknowns. The first three coefficients pertain to the three initial conditions and can be solved readily as $\mathbf{a_0 = 0}$, $\mathbf{a_1 = 0}$, $\mathbf{a_2 = u_t/2b}$. The final three, $\mathbf{a_3\ a_4\ a_5}$, can be found numerically using for instance the techniques of *Appendix A*.

The aforementioned calculation of a sequence θ_1/θ_2 at equal intervals of the independent variable α can also be achieved with a trajectory plan specified in terms of the independent variable $\mathbf{xo}$ since $\mathbf{u = b\ yo''(xo)/[1+yo'^2(xo)]^{3/2}}$ and $\mathbf{R\Delta\alpha = [1+yo'^2(xo)]^{1/2}\ \Delta xo}$.

CSM

How can we get by, achieving the millimeter level of positioning of **http://www.nd.edu/~sskaar/**, by employing nominal kinematics as freely as indicated above.

The trick is a subtle one: all of the above calculations involving the nominal forward kinematics of the arm as well as estimates of target location relative to the base are filtered through the same kind of camera-space interpretation described in

Chapter 7. As described in that chapter, the reference frame relative to which "physical position" estimates are based is in reality one that has a special interpretation: It is that particular Cartesian frame relative to which the current location of the end member would be properly described by the nominal kinematic model of that holonomic arm. And it shifts a bit as the region of the workspace within which end-member movement occurs is shifted. This is a frame that in position and orientation is somewhat different from the nominally fixed coordinate system with respect to which the kinematic model actually used in the calculations is described.

Note that both the holonomic arm and each of the two cameras sit atop and move with the mobile base. They move as one member, fixed together by the rigid base. Because of this, estimation of "view parameters" $\mathbf{C}_1$-$\mathbf{C}_6$ can be established in a manner identical to that of a stationary holonomic arm as described in Chapter 7. Any current camera-space kinematics estimate can serve as a camera-specific observation equation similar to those applied in Chapter 10 with the extended Kalman Filter algorithm. That is, current camera-space-kinematics estimates relate, by way of the flattening process, the appearance in camera space ($\mathbf{x}_c$ $\mathbf{y}_c$) of any one of cues **A**, **B** or **C** to the nominal location of the target trio - which can be condensed to x_B y_B z_B ϕ – relative to the mobile base's reference frame. And, as per the discussion of Chapter 7, the observations relate the target's position relative to the actual location x_B y_B z_B ϕ of an unknown reference frame that is close to, but different from, the known, mobile-base-fixed reference frame **x y z**.

We proceed as if this latter, actual reference frame is indeed the same as the frame **x y z** with which the above trajectory plans are updated. That is, we ignore for purposes of trajectory planning the disparity due to error in the local on-board kinematics' characterization between the true x_B y_B z_B ϕ and our estimated x_B y_B z_B ϕ at any juncture. Thus the corresponding state equations for this EKF are Equations 11.1.

As the robot nears its terminus, and the estimates of x_B y_B z_B ϕ become refined, an interesting thing happens with this scheme. First it should be noted that estimates of z_B translate directly to θ_3 and θ_4 of the on-board, holonomic arm and hence η. (Two angles are found from one number z_B because this elevation combines with the *a priori* requirement of orientation parallel to the ground.) Executing the wheel control trajectory plan as if the offset between frames did not exist results in the desired configuration in camera space between manipulated and target bodies. This may not be immediately intuitive, but it is easy to see in the limit of "arrival". If the bodies in question do in fact achieve camera-space configurations consistent with maneuver objectives then samples of cues **A**, **B**, and **C** will indicate x_B y_B z_B ϕ consistent with the target objectives and the nominal models. Leading up to that juncture it becomes a matter of whether the response to wheel movement of the base-fixed reference frame with respect to which x_B y_B z_B ϕ is actually estimated is close to the nonholonomic kinematics model for the base's nominal counterpart. It is easy to show that for small, primarily translational increments the answer is "yes".

CHAPTER 12

DESTINATION WITHIN REACH; NEW USES FOR THE INTERNET

Sometimes seemingly small shifts in a technological approach enable leaps in utility. Great inherent prospects of mechanical dexterity and mobility subject to computer control have long been suspected. Still, there must be that initial impetus, that initial demonstration that what we have is not just a laboratory toy, but something robust and valuable - directly adaptable to the real world. Creative and motivated users and entrepreneurs will take it from there. What are some forms of the possibilities?

Few analogies in technology are perfect, but machines that possess a general mechanical mobility and/or dexterity - robots - may finally come into their own in the same halting-then-surging way of computers - machines that possess the general ability to be programmed to execute digital algorithms. It is hard in this day and age to believe, but sixty years ago the electronic digital computer was thought to have very limited usefulness. The military could find uses in wartime - to calculate precisely the trajectory of a projectile, for example. But with peacetime, interest in the heat-generating, space occupying, vacuum-tube-consuming computers of the nineteen forties died temporarily.

Revival of interest came on slowly at first, but as cost and proficiency improved, people began to catch on, in stages. Business applications showed the devices could be used profitably and creatively. The extension of computing to individuals stirred further creativity, and demand, and supply, and use: spreadsheets, word processing, graphics.

Despite all kinds of effort – marketing, academic, and media – robots today seem stuck in a mostly teach-repeat mode, limited mostly to a few industries. Though possessing mechanical attributes that give them broad generality and usefulness-in-principle they find comparatively few profitable applications. Among individuals they are used mostly for entertainment.

With its self-correcting, three-dimensional robustness, ability to be supervised at a high, intuitive level by humans, and broad range of task prospects, we argue that camera-space manipulation can spur creative expansion of robot use comparable to historical periods of growth in computer use. Similarly, the practicality of extending teach-repeat to indoor operation of wheeled navigation could, in the hands of creative entrepreneurs, result in domestic service robots that have for years been forecast. Controlling effectively mobile dexterity - an onboard arm transported by a mobile base - adds yet more to these two prospects.

Factory

Calibrated visual guidance of robots did not deliver the workerless automobile factory in the 1980s as forecast. So human workers were retained – often because of their own ability to guide visually machines that might bear the weight of a tool, or a part to be added to an assembly. The ability of camera-space manipulation similarly to employ estimation over the course of any current maneuver to converge parts to be joined in two or more separated visual reference frames has been discussed herein. And this prospect for obviating the problems with *calibrated* artificial-vision-based guidance of robots of the 1980s experiments has been put forth. But perhaps a deeper CSM prospect can be appreciated by thinking about less-publicized attempts to create a different sort of workerless factory.

Gathering dust in mammoth rooms throughout the U.S. are elaborate webs of machinery designed to fully automate factory production without the use of any vision at all. "Hard automation" - specific, dedicated hardware to move components into known, fixed locations at all stages of production - has been tried in a range of industries from textiles to paper containers. Fixing product components allows a fixed, low-degree-of-freedom movement of machine tools and other bodies to interact with the product in a prescribed and uniform way, a clear virtue in the high-precision business of manufacturing. In fact this same philosophy has also been applied to automated storage and retrieval.

But the real-world problems that arise are real indeed. Reliability of fixed-movement operations must be ensured from start to finish. Conveyance, fixturing, machining, assembly and packaging become unforgivingly restricted. As parts wear out, anywhere throughout the process, specialized replacements are required right away or operation may halt. There is little opportunity to modify process or product. Wear that produces imperfection upstream can disable subsequent operation downstream. Jamming, clogging, all kinds of disruption cannot be precluded with a simple accommodative shift of positioning as with a human worker guided by his or her vision.

Teach-and-repeat robotics only alleviates a portion of this problem. It allows for the application of the more general-purpose, more versatile robotic arm. But the still-special-purpose need for active fixturing of workpieces remains. A simple repeating robot action has no ability at all to exploit the mechanism's inherent versatility to adjust motion to accommodate shifts in workpieces as presented, not even slightly. If calibrated vision is brought in the burden shifts from ensuring fixturing precision to ensuring precision of camera and robot-kinematics calibration. The cure is as bad as the disease.

Visual servoing, with its control in the sensors' frames of reference, seems like a robust answer. But it requires real-time feedback of a type that cannot be delivered in practice, for most applications. And it fails to exploit fully some powerful aspects of machine/computer systems such as the ability to register in memory and recall exactly recent samples of joint rotations and images, the ability to move according to precise mathematical instructions, or the time-tested, highly refined joint-level servomechanisms of robots.

Yet there is something clearly powerful about controlling the relative positions of two bodies in the reference frames of visual sensors. Humans can thread a needle this way, for example, a task that uses only vision for guidance.

So the writers have focused on the use of versatile mechanical dexterity combined with the idea that extremely high precision can be achieved robustly, as with the totally uncalibrated human needle threader, by eschewing absolute or "world" frames of reference and pursuing maneuvers in the frames of reference of visual sensors. The continuous two-dimensional sensor space of cameras is a domain that can and should be exploited for defining and pursuing maneuver objectives. Well-behaved, predictable asymptotic limits of real-world lens mappings and robot kinematics can be exploited to apply the immediate-past, current-maneuver record of approach to fine tune robot instruction for precisely correct maneuver termination. The redundancy in such information can be used as a reliable indicator of anticipated maneuver outcome and/or system health or degradation. Widely separated cameras, each with its own geometric advantage, can be applied to the maneuver outcome, bringing further helpful redundancy as well. As wear and heat produce their inevitable shifting effects on the hardware, CSM-guided robots compensate automatically.

Unlike humans who require simultaneous visual apprehension of both of the objects to be joined or otherwise relatively positioned, artificial systems can register the workpiece presence in camera space well ahead of introduction of the mechanical arm into each camera's field of view. In fact, multiple images acquired from a stationary camera, each with a different illumination pattern of laser spots, can be applied to the understanding of camera-space objectives. Correlation or matching of these spots among all participant cameras can be an additional, powerful tool in assessing camera-space maneuver requirements. The ability to apply multiple images under various assistive lighting conditions such as laser spots for workpiece-target identification, as well as the ability to tolerate complete obscuration of the workpiece by the end-member operation itself, are great attributes not enjoyed by any human assembly-line worker.

Pixel resolution limitations of cameras can be offset by zooming in optically to the workpiece or otherwise narrowing the cameras' fields of view.

The principal limitations of the method lie with the practical requirements of obtaining adequately precise camera-space kinematics estimates, outlined in Chapter 7, together with the need to determine camera-space objectives from image information. Where automatic image analysis to determine camera-space objectives is too difficult, or where human judgment is needed, the patented point-and-click surface point specification, for example to locate hole-drilling junctures for aircraft rivets, can be employed by a human supervisor. The internet with its worldwide reach is an ideal tool if fully automatic target recognition and specification is impractical. Such remote supervision, with its great outsourcing potential, would not be possible with direct human-in-the-loop control.

Early uses of CSM should occur with industrial tasks that are critical yet have no fully satisfactory alternative such as hazardous material handling, manipulation in a dangerous or uncomfortable environment, or applications like aircraft-rivet hole drilling where the scale is so large that fixture-based hard automation is impractical. Instances where disruption of existing operation will be minimal could also be among the first. So if a visually guided robotic system could be directly introduced to replace a human working in an undesirable job, leaving little to change before or after this stage of production, that could be attractive. Many tasks such as stacking and unstacking product at intermediate stages for pallet/forklift transport fit this description. See *Figure 12-1.*

Figure 12-1. CSM used for pallet stacking for forklift transport between stages of production. Note visual cue on end member.

Longer term, CSM's combination of versatility, precision and automatic status-monitoring might result in a reach comparable to that which computers achieved in (say) automatic inventory and supply management. The technology could become integral and fundamental to much of manufacturing – the first resort for any automation due to considerations of cost, robustness, and quality. Avoidance of specialized fixturing with its costly need for actively precise pre-positioning of workpieces, robustness stemming from the ability to adjust motion automatically in response to any wear or other geometry-altering transitions, and versatility of production could drive such an alternative mindset.

Getting around

Testing of the automatically guided wheelchair (patent owned by Notre Dame and the U.S. Dept of Veterans Affairs) of Chapter 10 occurred in a one-bedroom, narrow-corridor mobile home kept by the U.S. Dept. of Veterans Affairs at the Hines VA Hospital near Chicago. To pass the time the seated tester read a book over the course of being transported from position to commanded position within the home, eight hours in all. This freedom from having to pay attention to anything - other than an initializing menu selection of the next choice among 9 possible trailer destinations such as "kitchen sink" or "bed" – was striking. Since the chair knew the way, there was no need for the rider to have prior familiarity with the floor plan of the trailer. It was as if someone had built a special track in the floor for the chair to follow for every one of the taught starting-pose/end-pose permutations. But unlike iron tracks, these permit pure pivoting as well as multiple direction changes to maneuver in tight quarters.

With today's five hundred terabyte hard drives the potential number of such dedicated "tracks" is virtually unlimited (the storage requirement for even the longest trailer path is tens of kilobytes). And once the walls are treated with adequate cue coverage the time taken to create a new track is no more than several minutes for a human teacher who knows the way.

Moreover, if onboard ultrasound sensors detect an unforeseen obstacle en route along any one of the "tracks", the system can identically reverse course - backing up through path segments where it had gone forward to first reach the obstacle - to return to the original point and wheelchair-pose of departure. From there the human rider (or, easily enough, a built-in machine logic) could route the wheelchair to an alternative destination from which the original destination could be reached without nearing the obstructed part of the floor. It would be as if a traveler from South Bend to Indianapolis rerouted through Gary to avoid congestion in Kokomo. Better still, the cost of creating the indoor counterpart to a new Kokomo bypass would, in the trailer or within any cue-endowed building, be low enough to appeal even to Indiana legislators.

Institutions as well as homes often have a particular need to transport individuals who may be unable to find their own way visually. Outfitting buildings with wall cues or their functional equivalent allows the teaching of numbers of paths that might run into the thousands. Following suitable standardization this could permit even newcomers to download a disk of tracks into their own wheelchair and select their sequence of highly-specific destinations from a menu. Once the building is so outfitted adding other autonomous-vehicle functionality, such as autonomous, taught-path floor maintenance becomes straightforward.

Inventory sorting

The Web site ***http://www.nd.edu/~amemicro/simulation/simulation.html*** presents an interactive demonstration of a point-and-click system that commands a six-degree-of-freedom robot to pick up randomly arranged boxes and stack one selected box on top of another. *Figure 12-2* illustrates an image presented to the user's monitor. The user is asked to place the cursor above the first box to be engaged. One of two possible selections is illustrated in *Figure 12-2*.

Figure 12-2. The user is invited to point and click onto the top of the box that is first to be engaged by the robot.

Following point and click a site video shows a laser spot converging onto the user-selected juncture using the Jacobian as described in *Appendix D* and illustrated in *Figure 12-3*. The laser pointer blinks on and off synchronized with all three participant cameras' image acquisitions. The result is a common point location of the user-selected juncture as it appears in all three participant cameras. Two small, automatic additional moves of the pan/tilt unit allow for common location in all participant cameras of two additional junctures on the same user-selected surface of the box to be engaged.

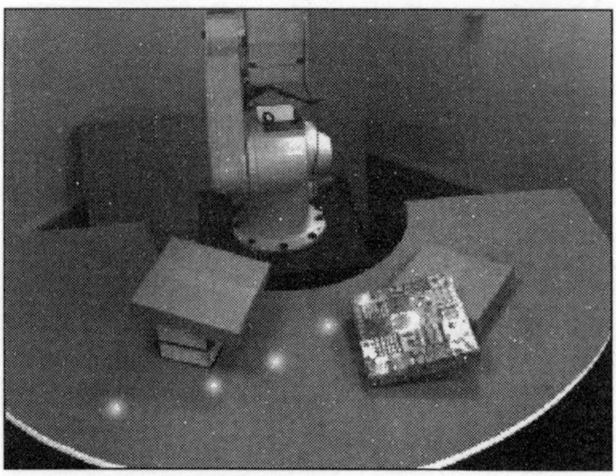

Figure 12-3. After clicking, a pan/tilt unit converges the laser spot to the selected juncture automatically.

This is followed by illumination with a multiple-spot laser pointer from the same pan/tilt unit that directs the initial spot as per *Figure 12-3*. As indicated in *Figure 12-4*, these spots are detected in all three cameras simultaneously. The spots are not readily apparent on the selected box, but the computer has no trouble finding them because of the application of image differencing. Subtracting the grayscale values of an image with the multiple-beam laser pointer turned off from each of three counterparts in *Figure 12-4* results in a clear location for each spot despite the visual complexity of the surface of interest.

Figure 12-4. From the multiple-beam laser pointer laser spots are detected readily in all three cameras using image differencing.

These simultaneous spots are then "matched" among the three cameras using the previous matches of the three individual spots. Matching is based on a roughly linear mapping between cameras of spot locations that fall on the same planar surface. Surface discontinuity is identified in part by using the fact that this roughly linear mapping between cameras of same-surface points ends at the discontinuity.

The separation between consecutive spots in *Figure 12-4* is a fairly wide three inches or so. This allows for reasonably lax tolerances for matching the spots among cameras. In the end six different groups of laser spots are cast, each with a slightly altered pan/tilt setting, each with its own group of images as per *Figure 12-4*, and each allowing for matching. Aggregating these spots allows for a precise and reliable automatic finding of the edges in each camera as indicated in *Figure 12-5*.

Figure 12-5. The dense spots, once matched, allow for the automatic specification of the boundary of the top surface and the camera-space locations of targets for the manipulator.

With camera-space targets computed robustly through the application of the laser spots the cue-bearing grasper enters the picture as indicated in *Figure 12-6*. Each of the three cameras acquires images en route to the terminus, refining the camera-space kinematics according to camera-space locations of the prominent cues located on the end effector. The result is precision of positioning well within the three millimeter tolerance of the grasper for box engagement. The Kawasaki Js5 robot enters the fields of view of the cameras and removes the designated box.

The Web site goes on to ask the user to point and click for a second box, which will be the target over which the Js5 will align the box currently in its grasp. The user may then select and align the remaining box leaving a stack as indicated in *Figure 12-7*.

The practical, real-world prospects for a similar kind of arrangement, applying any number of user-presented views and angles, are great. Operation is inherently robust due to the camera-space mode of control combined with exploitation of many kinds of imaging redundancy to ensure no surprises at the end of any maneuver.

Figure 12-6. Cues are detected en route to enable precise engagement. Obscuration of the target box by the grasper, a problem for human visual guidance, is not a problem where fixed-camera-space targets have been established in advance.

Figure 12-7. Following a second box placement atop the stack, the interactive demonstration leaves a neat alignment of three boxes. Human decision making is responsible for the order; but all robot motion is fully autonomous.

There are many instances of inventory sorting of this kind. And the dexterous machine is ideal for carrying these out with only the relatively high level of human supervision described.

Point and click your way to a new home

How about sitting in front of a computer in your old house's living room while on-site robots build you a new house? Okay, maybe that's a little exaggerated. But what in the end does building a house actually consist of? Much of it is positioning and moving objects relative to other objects. Accuracy and expertise of this action is required at nearly every step, but it is interesting that, with the assistance of a few comparatively simple visual-reference tools such as levels, plumb lines and laser lines, the reference frames within which the positioning is achieved are those uncalibrated, personal reference frames of human workers' vision. Provided the right kinds of mechanical dexterity and mobility are introduced, these same operations might be done if anything *more* conveniently in reference frames of judiciously positioned cameras placed throughout the construction site. And CSM permits pan/tilt/zoom motion by the remote supervisor of cameras to isolate the region of each maneuver in at least two cameras.

As the house becomes enclosed cameras and laser-pointer-bearing pan/tilt units need to be located inside, but this very location could be achieved prior to full house enclosure by the same mechanisms that finish the enclosing. Precise location of the cameras never matters any more than the precise location of a carpenter's eyes matter as he finishes off his creation. In fact, CSM's ability to register camera-space objectives well ahead of the introduction of the robot-borne tool, provided all participant cameras remain stationary from this target registration through maneuver execution, yields a big advantage that the human carpenter doesn't have: The introduced tool and end effector can entirely obscure the target body from participant-camera view throughout maneuver execution with no loss of accuracy.

Other than automatic positioning the main requirement is judgment and expertise. In what order, and where, will the positioning events occur? Answers to these questions would be supplied by a human supervisor, possibly a remote supervisor located anywhere within the internet's reach.

Some tasks such as hole drilling or the positioning of a nail gun would be relatively straightforward as described in Chapter 7. Tasks involving the positioning of large boards to be fastened into place would be more complex, but can be handled by visually assessing, after a board has been engaged and grasped, the geometry of the relationship between the board's perimeter boundaries and the cue-bearing robot end effector. Such assessment is straightforward and precise using multiple laser spots of Chapter 8, and it permits maneuver referencing off of cues located on the robot's grasping mechanism.

Many other categories of application, including farming, surface preparation, maintenance, and ordnance handling are also within reach. As our vision extends beyond tasks and task sequences that have been evolved specifically to allow for human execution, the range of prospects may grow far greater.

Moving targets

Non-stationary targets may be roughly divided into two categories: Targets that move in response to the action of the robot and targets that have their own trajectories independent of that action. The baseball-fielding example of the Foreword is one instance of the latter, as is the ping-pong-ball-catching video which can be found in **http://www.nd.edu/~sskaar/**.

This video shows first the catching event itself, and follows on with an indication of the CSM strategy that can be adopted to handle moving objects. Early glimpses of the ball in flight are registered in the camera space within which the catching maneuver is controlled (just one camera in this case because the flight of the ball is restricted to a plane.) A very simple model, one that actually ignores air drag – for a ping-pong ball quite a significant effect – is sufficient to allow for a prediction of the flight of the ball to improve on a timetable sufficient to permit the system to place the end member, in that same camera space, in time for projectile interception. The main point here is that just as artificial systems can use/recall quantitative data pertaining to stationary objects so too they can apply quantitative data and specific physical models to project roughly, but in many cases adequately, the camera-space future of an unfolding event.

If movement of the target object represents a shifting, rather than a primarily elastic-deformation, response to contact with the robot, then many of these techniques are, to the extent of any such shift, invalidated. The advantage of CSM is that it can apply several video samples of the target object within several different cameras under varying laser-spot and other contrived lighting situations – including multiple images under varied lighting from any one camera – to be applied to the manipulation event. All this assumes that the target stays put in camera space. Cameras and target objects must remain stationary (unless, as per the ball-catching example, the camera-space trajectory of the target can be forecast.)

CSM logic

There is a logic for applying distant, separated, and stationary cameras to the robot-control problem. And there is an additional logic for applying camera-space kinematics, estimated and refined as the robot approaches a terminus, where the distant cameras supply the reference frames for maneuver control.

The starting point of the logic is the premise of Chapter 6, repeated at the beginning of Chapter 7: The collocation in each two-dimensional camera space of two or more well-separated cameras of a chosen point that moves as a conceptual 3D rigid-body extension of the robot end member with a particular stationary-surface point, previously registered in these same stationary camera spaces, guarantees physical, three-dimensional collocation of the manipulated and stationary points. Though already nearly tautological, this premise can be verified experimentally.

Consider the drill apparatus of Chapters 7 and 8 as it is presented in *Figure 12-8.* Note that a cue has been located very near to the drill-bit tip. Due to the close proximity of this cue's center with "Point P" – the entry juncture on the drill bit – it is easy to apply appearances of the cue to refine very precisely and accurately camera-space kinematics of the drill tip in the local joint-space vicinity of contact/entry rela-

Figure 12-8. Near-locally-perfect camera-space kinematics can be realized for the drill tip if a cue is located near that tip during target-cue approach. Established in 2 or more cameras, this camera-space-prediction perfection results in virtually perfect 3D positioning of the tip at the target's center. That is within about two thirds of a pixel in each camera. But how good a job estimating the camera-space kinematics of the positioned drill tip can we do if restricted to samples of the cues shown well above this tip?

tive to the stationary "target" cue. Other cues moving with the end member may be used to estimate the camera-space kinematics but strong weighting of the tip cue in two or three poses close to entry leads to a high degree of local confidence for that point.

Of course the tip cue must be removed prior to actual drill contact with the target cue, but this need only occur *slightly* before contact. Consistently, when this is tried, that contact results as predicted with respect to the stationary cue's center: Physical error in three-dimensional space is under one or two millimeters - or about half of one pixel in each of the three participant CSM cameras. Better physical precision would follow higher camera resolution.

The logic develops further with the question: how good a job can we do achieving accurate camera-space kinematics for a positioned point – say the tip of the drill bit - if we can only make use, during approach, of cues that are some distance away from this juncture on the end member? This is the question of *Figure 7-14*. And, as suggested by that figure, it is a question that ultimately must be verified experimentally. For the system of *Figure 12-8* reliance for tip positioning only on end-member cues shown some distance away from that tip only increases positioning error by perhaps a factor of two – to a maximum of around 3mm. What factors influence these comparisons?

The underlying issue has to do with the goodness, or accuracy, of the asymptotic limits implicit in our model as developed in Chapter 7. This accuracy, in turn, relates to many realities of the system elements including: optics, mechanism kinematics, and cue locations - cue locations relative to each other on the end effector and relative to the positioned point(s). The accuracy also relates to sampling frequency, sampling position and sampling quality of the visually accessible cues during approach.

To appreciate these elements it is worth making a distinction between two kinds of camera-space-kinematics-estimation error: namely random error vs. model error. Random error is considered for our application as principally effects associated with the pixel quantization of camera space, which like many things is actually deterministic but with attributes that allow us to treat it as random. (You can't for any particular extent of pixel resolution infinitely resolve the camera-space location of cues, be they the circular ring cues or the laser spots.) Importantly, in contrast with calibration, the CSM framework allows us to treat this "random" error, legitimately, as "zero-mean".

Deterministic model error associated with imperfection of Chapter 7's asymptotic limit includes: the departure locally of the optics from the pinhole ideal; error in assessing Chapter 7 $\mathbf{Z_o}$; and departure from nominal of the robot's kinematics. There are many steps pertaining to lens selection and/or cue positioning that inexpensively but reliably draw realized three-dimensional positioning error close to zero, given the CSM approach.

In general the "antidote" to *random* error - insofar as the camera-space kinematics are concerned - is more visual samples en route to the terminus, in an effort to achieve a kind of averaging effect. Not only is Moore's law in our favor here, but the requisite burden of timely image analysis (it must be completed prior to maneuver termination for any new sample to help) can relatively easily be distributed among several parallel processors. The antidote to deterministic error is improving and expanding the region within which Chapter 7's asymptotic limits are valid. An alternative tack is to skew sample weighting more highly in favor of samples sufficiently close to the terminus to be more nearly consistent with those asymptotic limits. Suffice it to say that the real-world options for achieving precision objectives are generally affordable and varied.

Just as the application of more end-member cue samples during approach can overcome pixel-quantization and the "random error" that it causes with the camera-space kinematics, so too more pre-approach samples of the workpiece can overcome pixel-quantization limitations on the target side. This is part of the advantage of stationary cameras and pan/tilt-directed laser spots. The "matching" among cameras of densely packed laser spots across workpiece surfaces requires significant separation among the individual spots in a given image. But multiple images, with slight pan/tilt-caused laser-spot offsets in each, produce a dense packing of the matched spots (such as *Figure 12-5* above) which can in turn be applied to highly accurate best fits of the surface contour. Because of the application of camera-space manipulation to address that surface with the robot end member, nominal Cartesian (x-y-z) coordinates of junctures on this best-fit surface are perfectly compatible (locally) with the forward-kinematics model of the robot. The result is robust subpixel tool-tip delivery.

"Tool-tip delivery" should, in this context, be interpreted broadly. There are very few practical tasks - even highly elaborate tasks - that cannot be construed in terms of a predetermined redefinition/shift of "Point P" on the end member as the addressed juncture on the best-fit surface evolves. When combined with surface-referenced orientation information and control, as for example with the cutting task of *Figures. 8-10, -11, -12, -13*, robot movement can be realized for a broad range of task descriptions.

Distant cameras are unaffected by dirt, debris and movement in the vicinity of the robot action. They can be panned, tilted and/or zoomed before coming to rest to lend their guidance across a large swath of potential workspace for one or several robots. This region-enlarging, resolution-enhancing resetting of cameras does not come at the expense of loss of applicability of samples from a parameter-initializing, preplanned-motion event. Information contained in the first four elements of the pre-pan/tilt/zoom camera movement can be applied after that movement. When combined with new post-movement samples of end-member cues and associated robot-joint angles, all six elements of $\underline{\mathbf{C}}$ quickly return to their former accuracy but in accordance with the new pointing/zoom condition.

Robots as well need not be stationary, since no calibration is involved; and cameras may either move with the robot base or be located on fixed platforms so as to be able, in sufficient numbers, to access visually a given robot's target object surface. The potential for multiple redundancy in terms of numbers of cameras directed toward completion of any given task is likewise advantageous. Part of the reason has to do with the enhanced averaging effect of added cameras when robot joint rotations are resolved for robot-servomechanism command. But part also has to do with the task/precision monitoring that having multiple cameras/observers enables. In particular, the *independence* of the several two-dimensional reference frames means that significant "agreement" regarding resolved robot-internal-pose commands among several such observers virtually ensures precision relative to designated junctures on the workpiece surface. The order of magnitude of this precision should statistically, provided the camera-view directions are well-distributed, be closely related to the minimized quantity $\mathbf{J}(\underline{\boldsymbol{\theta}})$, as described in Chapter 7. The self-correcting system, if it is determined that insufficient consistency among cameras' estimates for the current maneuver are in place, can "back off" the robotic arm - in order both to reacquire target data and recompute from a new, second approach the refined camera-specific view parameters. Rather than precision being enforced in the traditional way of calibrated hardware, it will instead be enforced with the combination of mechanical versatility and light. The promise of such a shift is vast.

The indoor control of autonomous wheeled robots may require, as a practical matter, the extension to teach/repeat of nonholonomic robots. Outside of the following of tracks on the floor, teach/repeat offers the one approach demonstrably, reliably able to deliver a wheeled vehicle from pose to pose within a home. It offers the flexibility not offered by track-following systems of convenient redefinition of new paths, ability to depart from and return to a path on a contingency basis (obstacle-avoidance for example), and multiple direction changes as well as pivoting such as may be needed in practice in small corridors or rooms.

The inexpensive installation of wall-mounted cues enables teach/repeat control to be exercised with all kinds of wheeled vehicles. In public places and institutions this should translate into multiple use. Floor maintenance, patrolling, wheelchair guidance, mail delivery and so on. When combined with a programmed-in logic to shift en route to alternative path segments, the teach/repeat paradigm permits all kinds of responsive versatility based upon sensing. Proximity-sensor-based departure from a path, for example, possibly to join an alternative, previously taught path segment, or rejoin the original path, should be straightforward to program into such a system. Moreover, there is no reason that such an automatically created new trajectory should not be automatically added to the library of stored paths.

This kind of system learning should be practicable and might lead to systems that appear to have real intelligence. Importantly, however, systems would always be grounded in the apriori cue information.

Detection of wall cues combined with wheel odometry represents multiply redundant incoming information. As with CSM above, such redundancy, since it is internally consistent, translates into statistical certainty. As is also true of CSM, more cameras - in this case looking outward from the vehicle – produce enhanced precision as well as higher levels of such statistical certainty pertaining to current-position estimates.

Finally, the marriage of the two technologies – CSM on the one hand, and the autonomous navigation of extending teach/repeat to indoor, wheeled vehicles on the other – has prospect as well. Maintenance, for example, is not restricted to floors. Routine tasks such as wastebasket emptying or chalk-board erasing should be addressable with the combination of these two very different robot-control strategies: teach/repeat to locate the robot base within the building and CSM to deliver the end member of an arm as required to complete the manipulation part of the task. Such application of CSM could include the wheels of the base as needed degrees of freedom for producing the needed range of end-member position/orientation, in which case the mobile camera-space manipulation of Chapter 9 would be appropriate. Alternatively, the mobile-base-borne arm may have all the dexterity in its own right needed to complete the manipulation task at hand, in which case the role of the mobile base would be simply to transport the arm to within reach of each manipulation objective.

The imagination extends the list of possibilities, as always. What we suggest is that a shift of technique will open the flood gates of realistic utility. The hardware, we believe, is already here.

RELATED ARTICLES

1. ***Skaar SB:*** "Vision-Based Robotics Using Estimation," a multimedia monograph of ONR-sponsored research, WWW/Mosaic system (on the internet), *http://www.nd.edu/NDInfo/Research/sskaar/Home.html*

2. ***Skaar SB, Gonzalez-Galvan E:*** "Versatile and Precise Manipulation Using Vision," Teleoperation and Robotics in Space, Skaar, Steven B. and Carl F. Ruoff (editors), *AIAA*, Washington, D.C., pp. 241-279, 1994.

3. ***Skaar SB, Yoder JD:*** "Extending Teach-Repeat to Nonholonomic Robots," *Smart Structures, Devices and Systems*, Prentice Hall.

4. ***Skaar SB, Brockman WH, Hanson R:*** "Camera Space Manipulation," *International Journal of Robotics Research*, Vol. 6, No. 4, pp. 20-32, Winter 1987.

5. ***Skaar SB, Brockman WH, Jang WS:*** "Three Dimensional Camera-Space Manipulation," *International Journal of Robotics Research*, Vol. 9, No. 4, pp. 22-39, August 1990.

6. ***Skaar SB, Yalda-Mooshabad I, Brockman WH:*** "Nonholonomic Camera-Space Manipulation," *IEEE Transactions on Robotics and Automation*, Vol. 8, No. 4, pp. 464-479, August 1992.

7. ***Tang L, Skaar SB:*** "Stability of Conventional Controller Design for Flexible Manipulators," *ASME Journal of Applied Mechanics*, Vol. 60, No. 2., pp. 491-497, June, 1993.

8. ***Chen WZ, Korde U, Skaar SB:*** "Position-Control Experiments Using Vision," *International Journal of Robotics Research*, Vol. 13, No. 3, pp. 199-208, June 1994.

9. ***Baumgartner ET, Skaar SB:*** "An Autonomous, Vision-Based Mobile Robot," *IEEE Transactions on Automatic Control*, Vol. 39, No. 3, pp. 493-502, March 1994.

10. ***Miller RK, Stewart DG, Brockman WH, Skaar SB:*** "A Camera Space Control System for an Automated Forklift," *IEEE Transactions on Robotics and Automation*, Vol. 10, No. 5, October 1994, pp. 710-716.

11. ***Yoder JD, Baumgartner ET, Skaar SB:*** "Initial Results in the Development of a Guidance System for a Powered Wheelchair," *Transactions on Rehabilitation Engineering*, Vol. 4, No. 3, September 1996, pp. 143-151

12. ***Gonzalez-Galvan E, Skaar SB, Korde UA, Chen WZ:*** "Application of a Precision Enhancing Measure in 3-D Rigid-Body Positioning Using Camera-Space Manipulation," *International Journal of Robotics Research*, Vol. 16, No. 2, April 1997, pp. 240-257

13. ***Yoder JD, Skaar SB, Arriola H:*** "Using Probability Estimates to Identify Environmental Features for a Nonholonomic Control System", *AIAA Journal of Guidance, Control and Dynamics*, Vol. 20, No. 6, November-December, 1997, pp. 1215-1221.

14. ***Seelinger M, Gonzalez-Galvan EJ, Robinson M, Skaar S.B:*** "Robotic Plasma Spraying Operation Using Vision", special issue of the *IEEE Robotics and Automation Magazine* on "Applied Visual Servoing", Vol. 5, No. 4, Dec. 1998, pp. 33-38.

15. ***Lin T D, Skaar SB, O'Gallagher J:*** "Proposed Remote Control/Solar Power Concrete Production Experiment on the Moon", *ASCE Journal of Aerospace Engineering*, Vol. 10, No. 2, April 1998, pp. 104-109.

16. ***Gonzalez-Galvan EJ, Skaar S.B, Seelinger M:*** "Efficient camera-space target disposition in a matrix of moments structure using camera-space manipulation", provisionally accepted for appearance in the *International J. of Robotics Research*, 1999.

17. ***Gonzalez-Galvan EJ, Pazos-Flores F, Skaar SB Cardenas-Galindo A:*** "Camera Pan/Tilt to Eliminate the Workspace Size-Pixel-Resolution Tradeoff with Camera-Space Manipulation", *Robotics and Computer-Integrated Manufacturing*, 18(2) pp. 95-104, April 2002, Elsevier Science Press.

18. ***Seelinger M, Yoder JD, Baumgartner ET, Skaar SB:*** "High Precision Visual Control of Mobile Manipulators", *IEEE Transactions on Robotics and Automation*, Vol. 18, No. 6, pp. 957-965, December 2002.

19. ***Cardenas A, Seelinger M, Goodwine B, Skaar SB:*** "Vision-Based Control of a Mobile Base and On-Board Arm," *Int. Journal Robotics Research*, Volume 22, No. 9, September 2003.

20. ***Skaar SB, Del Castillo G, Fehr L:*** "Extending Teach and Repeat to Pivoting Wheelchairs" *Systemics, Cybernetics and Informatics (JSCI)*, Volume 1, No 1.

APPENDIX A

FINDING LOCAL MINIMA OF A WEIGHTED LEAST-SQUARES FUNCTION

Many times a least-squares solution is sought in an attempt, for example, to fit data to a model. One common formulation entails minimization of a function $J(\underline{\boldsymbol{\beta}})$ having the general form

$$J(\boldsymbol{\beta}) = \tfrac{1}{2}\ \underline{\mathbf{r}}^T [W]\ \underline{\mathbf{r}} = \tfrac{1}{2} \Sigma_i W_i r_i^2, \quad i=1, 2, 3 \ldots n$$

where $\mathbf{W}_i$ is a "weight" assigned to each of n samples z_i, i=1, 2, 3 ... n (and where by extension the **nxn** matrix **[W]** is diagonal and therefore symmetric with nonzero, diagonal elements comprised of these numbers), and where each of the n "residuals" are given by $r_i = z_i - h_i(\underline{\boldsymbol{\beta}})$. The function h_i depends on the functional form associated with the model. Elements of $\underline{\boldsymbol{\beta}} = [\underline{\boldsymbol{\beta}}_1\ \underline{\boldsymbol{\beta}}_2 \ldots \underline{\boldsymbol{\beta}}_m]^T$ are the m parameters to be estimated from the data. In general $m<n$.

As an example, consider the function of *Figure A-1.* The functional form here is $\mathbf{y(x)=sin(\Omega x+\phi)}$, where in this case we will attempt to converge onto true values of Ω =1.6 and ϕ =1.4. Suppose that we were given the following three data points from this function:

y(1.1) = -0.018406
y(1.3) = -0.331985
y(2.0) = -0.993691

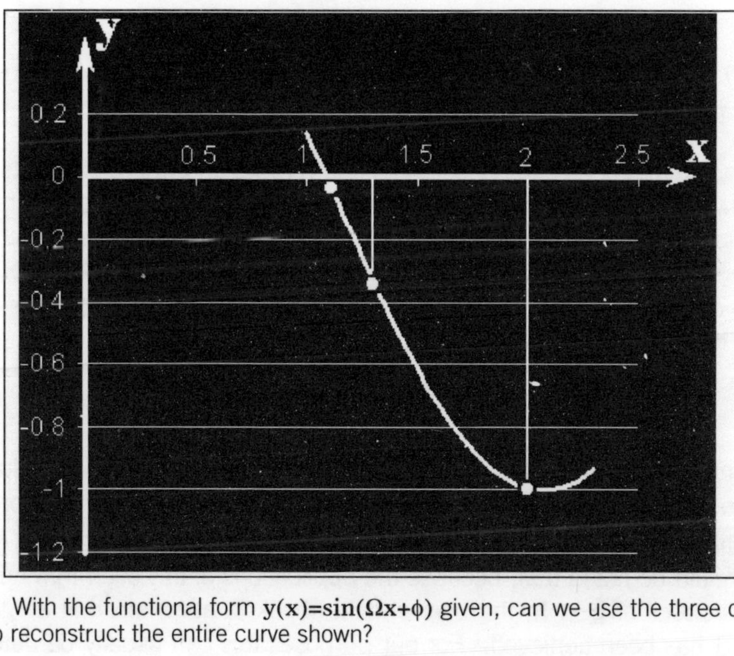

Figure A-1. With the functional form $\mathbf{y(x)=sin(\Omega x+\phi)}$ given, can we use the three data points indicated to reconstruct the entire curve shown?

These points are indicated on the plot of *Figure A-1*. Our task is to solve for Ω and ϕ. Referring to the formulation above, we identify

$$z_1 = -0.018406$$
$$z_2 = -0.331985$$
$$z_3 = -0.993691$$
$$h_1 = \sin(1.1\beta_1+\beta_2)$$
$$h_2 = \sin(1.3\beta_1+\beta_2)$$
$$h_3 = \sin(2.0\beta_1+\beta_2)$$

where we have named $\beta_1=\Omega$ and $\beta_2=\phi$. With equal weighting, $W_1=W_2=W_3=1.0$,

$$J(\beta_1,\beta_2) = \tfrac{1}{2}\{[-0.018406-\sin(1.1\beta_1+\beta_2)]^2 + [-0.331985-\sin(1.3\beta_1+\beta_2)]^2 + [-0.993691-\sin(2.0\beta_1+\beta_2)]^2\}$$

Here, m=2 and n=3. The following algorithm is credited to the Swiss mathematician Euler and the German mathematician Gauss. Begin with a best guess, β_{1o} and β_{2o}, of the parameters. For example, let β_{1o}=1.5, and β_{2o}=1.5. Let the first improvement on these values be $\beta_1 = \beta_{1o} + \Delta\beta_1$ and $\beta_2 = \beta_{2o} + \Delta\beta_2$.

J can be expanded about $\beta_o = [\beta_{1o}\ \beta_{2o}\ \ldots\ \beta_{mo}]^T$ using the Taylor series, and rewritten:

$$J(\Delta\beta) = \tfrac{1}{2}\,\mathbf{r}^T(\beta_o+\Delta\beta)\,[W]\,\mathbf{r}\,(\beta_o+\Delta\beta) = \tfrac{1}{2}\,\{\mathbf{r}(\beta_o)+[A]\Delta\beta\}^T\,[W]\,\{\mathbf{r}(\beta_o)+[A]\Delta\beta\} + \text{H.O.T.}$$

Where the matrix $[A]$ has m columns and n rows, and where elements of $[A]$ are defined such that Aij is the partial derivative of hi with respect to β_j, evaluated at $\beta=\beta_o$. Neglecting the higher-order terms (H.O.T.), and recognizing the symmetry of $[W]$, application of the "necessary conditions" for minimizing J over all $\underline{\Delta\beta}$ - i.e. the partial derivative of J with respect to each element of $\underline{\Delta\beta} = [\Delta\beta_1\ \Delta\beta_2\ \ldots\ \Delta\beta_m]^T$ is zero - produces:

(A1) $$\Delta\beta = [A^TWA]^{-1}A^TW\mathbf{r}(\beta_o)$$

The improved estimate $\underline{\beta} = \underline{\beta_o} + \underline{\Delta\beta}$ gets redefined as Eq. A1 is reevaluated using the improved $\underline{\beta}$ as the new $\underline{\beta_o}$ to produce yet another improvement. This process continues until convergence, which may be defined as $\underline{\Delta\beta}^T\underline{\Delta\beta}<\varepsilon$, where is ε a small number. It should be noted that, because the applications of this algorithm entail nonlinear appearances of $\underline{\beta}$ in $\mathbf{h}$, the user should check to ascertain that a "global minimum" of J has been achieved. For our purposes this can usually be determined in software based upon physical considerations.

For our example, with the initial guess of $\beta_{1o}=1.5$, $\beta_{2o}=1.5$, it is a good exercise to show that the initial vector of residuals becomes $\mathbf{r}(\underline{\boldsymbol{\beta}}_o)=[-0.0100\ -0.0284\ -0.0162]^T$. Also, the elements of $[\mathbf{A}]$ are given by:

$$\begin{matrix} -1.1000 & -1.0000 \\ -1.2387 & -0.9528 \\ -0.4215 & -0.2108 \end{matrix}$$

The initial application of Equation A1 $\underline{\Delta\beta} = [0.0825\ -0.0793]^T$ results in an improvement given by

$$\beta_1 = \beta_{1o} + \Delta\beta_1 = 1.5+0.0825 = 1.5825$$

$$\beta_2 = \beta_{2o} + \Delta\beta_2 = 1.5-0.0793 = 1.4207$$

After three or four more iterations, the reader can verify that we reach the answer that in this illustration/simulation was known at the outset: $\beta_1=1.6000$ $\beta_2=1.4000$. These values, combined with the assumed functional form recreate exactly the function shown in *Figure A-1.* The percent improvement associated with each consecutive iteration generally improves as the relative size of the neglected higher-order terms above diminishes.

APPENDIX B

FORWARD KINEMATICS
MODELING OF POINTS ON AN END MEMBER

A common way, used for example by many robot manufacturers, to specify the nominal forward kinematics model for a robot makes use of a 4x4 "homogeneous transformation matrix" such as the one indicated in *Figure B-1*.

$$\begin{Bmatrix} x_0 \\ y_0 \\ z_0 \\ 1 \end{Bmatrix} = \begin{bmatrix} \cos\theta_1 & -\sin\theta_1 & 0 & 0 \\ \sin\theta_1 & \cos\theta_1 & 0 & 0 \\ 0 & 0 & 1 & 0 \\ 0 & 0 & 0 & 1 \end{bmatrix} \begin{Bmatrix} x_1 \\ y_1 \\ z_1 \\ 1 \end{Bmatrix} \quad \text{(B1)}$$

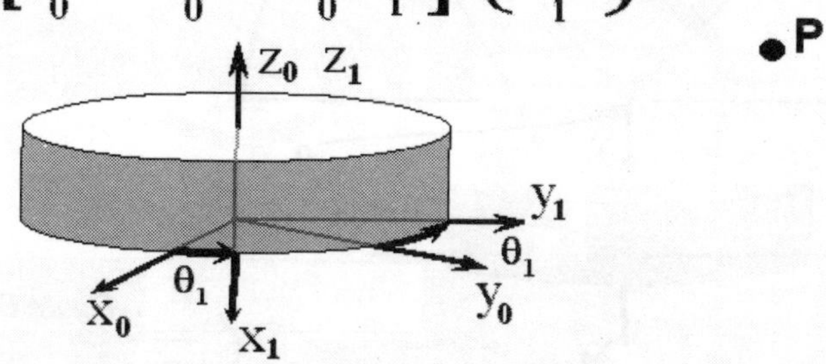

Figure B-1. This robot base entails a single rotation about the vertical axis. The "0" coordinate frame is stationary whereas the "1" frame rotates with the first angle θ_1 as indicated.

Consider an arbitrarily located point **P** as shown in *Figure B-1.* This point has coordinates at any moment that *could* be specified relative to the stationary "0" frame or relative to the "1" frame that rotates with the base. Let $[x_0\ y_0\ z_0]^T$ denote at any moment the coordinates relative to the stationary frame and $[x_1\ y_1\ z_1]^T$ denote the coordinates relative to the frame that rotates with the base about the vertical z_0/z_1 axis. Equation B1 uses a 4x4 homogeneous transformation matrix to specify how, at any moment, the two reference frames' coordinates of point P relate to one another. This relationship of course varies with θ_1. It should be noted that the upper 3x3 matrix embedded within the 4x4 matrix of Eq. B1 is the "direction cosine matrix". Any given element of that matrix is the cosine of the angle that separates the indicated axes. So for example the cosine of the angle that separates the positive x_0 axis from the positive y_1 axis is "$-\sin\theta_1$". The bottom row of the homogeneous transformation matrix as used here is always the indicated **0 0 0 1**. The top three elements of the rightmost column are all zero because the two reference frames in question share the same origin.

That is not the case for the next homogeneous transformation as we move from base to tip. Note that the origin o of the "1" coordinate system, and the origin o' of the "2" coordinate system are not the same point. They are separated by a vertical distance D. Specifically **o'** is located a distance D from **o** in the ***positive*** z_1 ***direction***. As a consequence the second homogeneous-transformation matrix, the one that relates **P**'s coordinates in the "2" frame to those in the "1" frame, has the following top four elements of its rightmost column: **0 0 D**. The reader can verify that the upper-left 3x3 matrix of the homogeneous transformation matrix of Eq. B2 represents direction cosines between indicated pairs of axes of the pertinent reference frames.

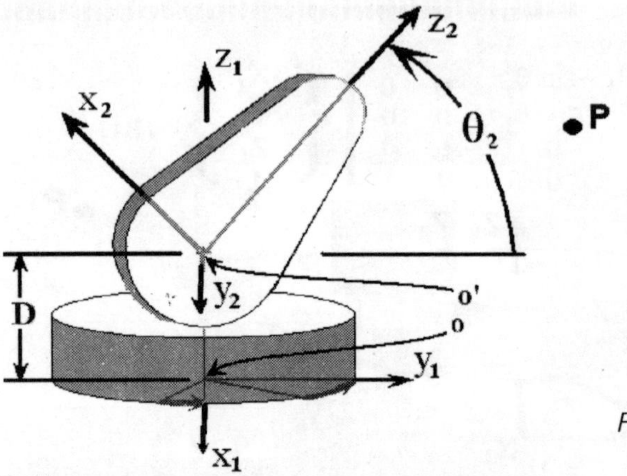

Figure B-2. Robot's second rotation.

$$\begin{Bmatrix} X_1 \\ Y_1 \\ Z_1 \\ 1 \end{Bmatrix} = \begin{bmatrix} 0 & 1 & 0 & 0 \\ -\sin\theta_2 & 0 & \cos\theta_2 & 0 \\ \cos\theta_2 & 0 & \sin\theta_2 & D \\ 0 & 0 & 0 & 1 \end{bmatrix} \begin{Bmatrix} X_2 \\ Y_2 \\ Z_2 \\ 1 \end{Bmatrix} \quad \textbf{(B2)}$$

The third angle of rotation θ_3 has defined for it a homogeneous transformation matrix as indicated in *Figure B-3*. In this instance o' and o" are separated along two components of direction; as referred to the "2" frame we have: **0**, **a** and $\mathbf{L_1}$, as shown. Note the ease with which Eq. B3 can be substituted back into Eq. B2, and the result from this multiplication can be substituted back into Eq. B1. This leaves a triple matrix product whose result is indicated in Eq. B4. For compactness we shorten, for example, $\cos\theta_3$ with c_3.

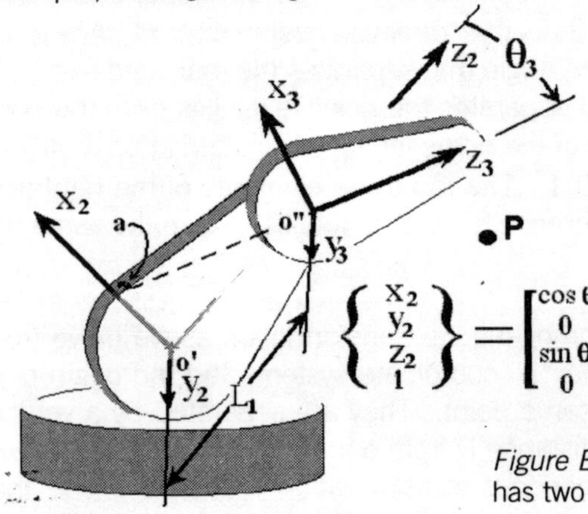

$$\begin{Bmatrix} X_2 \\ Y_2 \\ Z_2 \\ 1 \end{Bmatrix} = \begin{bmatrix} \cos\theta_3 & 0 & -\sin\theta_3 & 0 \\ 0 & 1 & 0 & a \\ \sin\theta_3 & 0 & \cos\theta_3 & L_1 \\ 0 & 0 & 0 & 1 \end{bmatrix} \begin{Bmatrix} X_3 \\ Y_3 \\ Z_3 \\ 1 \end{Bmatrix} \quad \textbf{(B3)}$$

Figure B-3. The third homogeneous transformation has two components of origin offset.

One big advantage of characterizing the large transformation matrix of Eq. B4 for any given robot is that it makes specification of the nominal forward-kinematics model, as required for estimating camera-space kinematics estimates in Chapter 7, straightforward. Consider the case where our point **P** happens to lie on and move with the x_3-z_3 plane as indicated in *Figure B-4*. The coordinates of this point are permanently $x_3 = 0$, $y_3 = 0$, $z_3 = L_2$.

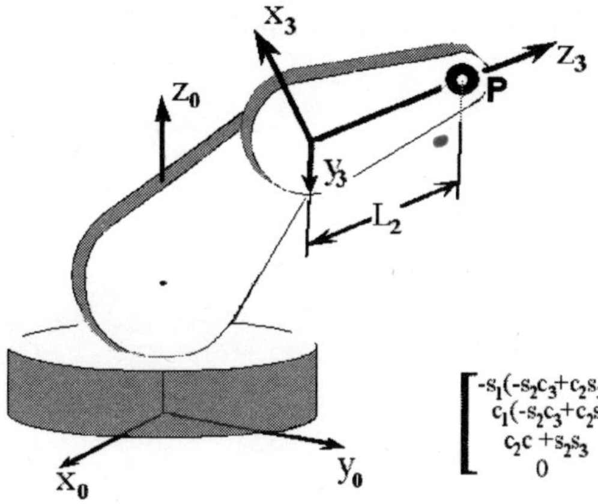

$$\begin{Bmatrix} x_0 \\ y_0 \\ z_0 \\ 1 \end{Bmatrix} = \begin{bmatrix} -s_1(-s_2c_3+c_2s_3) & c_1 & -s_1s_2s_3-s_1c_2c_3 & ac_1-L_1s_1c_2 \\ c_1(-s_2c_3+c_2s_3) & s_1 & c_1s_2s_3+c_1c_2c_3 & as_1+L_1c_1c_2 \\ c_2c+s_2s_3 & 0 & -c_2s_3+s_2c_3 & D+L_1s_2 \\ 0 & 0 & 0 & 1 \end{bmatrix} \begin{Bmatrix} x_3 \\ y_3 \\ z_3 \\ 1 \end{Bmatrix} \quad \textbf{(B4)}$$

Figure B-4. Relationship between the base and end-member coordinate systems.

Substitution of this into Eq. B4 yields the nominal forward kinematics model of the cue center denoted in Fig. B4 as point P:

$$x_0=-L_2s_1\,(s_2s_3+c_2c_3)+ac_1-L_1s_1c_2$$

(B5) $$y_0=L_2c_1\,(s_2s_3+c_2c_3)+as_1+L_1c_1c_2$$

$$z_0=L_2\,(-c_2s_3+s_2c_3)+D+L_1s_2$$

Most other holonomic robots can be similarly treated. Denavit-Hartenberg frame assignments are a natural starting place for defining the elements of the homogeneous transformation matrix though they are not used here. In Chapter 7 the end-member-referenced coordinates either of a cue or the positioned point are denoted by **X**, **Y**, **Z** - in place of x_3, y_3, z_3 used here. Such specification is exceptionally convenient and consistent for purposes of CSM when used with the homogeneous-transformation convention exemplified above.

The choice of definition of the intermediate reference frames is to some extent arbitrary. And the homogeneous transformations will appear differently depending upon this choice. Of course the final forward kinematics specification for any given point P should be the same regardless of this choice provided the same base frame is used. In the development above there is some difference in the form of the forward kinematics compared with the example used in Chapter 7 for the same robot model. Two reasons for this: A different base coordinate frame was used there (note that, as discussed in Chapter 7 it is best to apply the forward kinematics model to a frame whose origin is "near" the region where the maneuver is to terminate.) And the separation "**a**" of Equation B4 was taken for illustration to be zero in Chapter 7.

APPENDIX C

CIRCULAR/ELIPTICAL CUE DETECTION-THE SCANCUE ROUTINE

This section gives a detailed description of the SCANCUE algorithm. This routine reliably detects cues with a predefined shape in an image. It was designed to be a fast algorithm that returns the correct cue center information and that rejects *false positives*. False positives are cue-like artifacts that exist in the workplace: wall texture, objects with stripes, written characters on the wall, power outlets, etc. SCANCUE is able to reject those false readings without rejecting a real cue, which would be called a *false negative*, within certain preset limits.

This scanning routine performs a two-phase process: a fast detection to find cue centers in an image, followed by pattern matching techniques to reject false positives.

The first phase of SCANCUE was designed by Dr. Steven Skaar and Dr. Richard Miller. It searches quickly the possible cues in the image, and subsequently performs several checks to validate that the detected artifact is indeed a cue and not noise, texture, or some type of false positive identification. Two kinds of cues are employed, "white" and "black". A white cue is a black ring over a white background. A black cue is a white ring over a black background. They are both shown in *Figure C-1*.

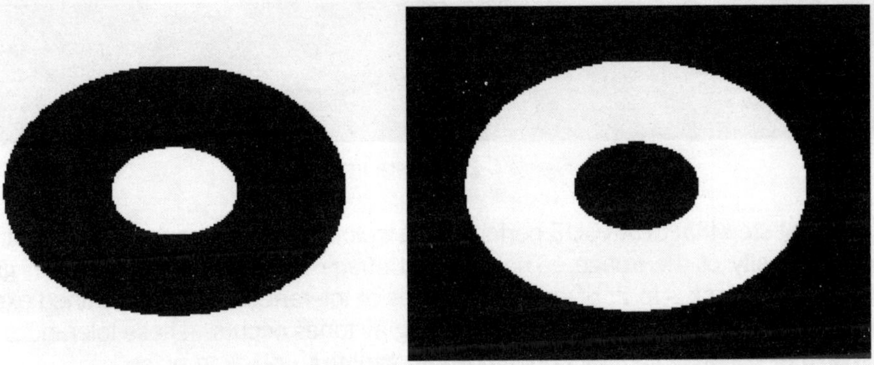

Figure C-1. Cue Types.

From a frame grabber device, SCANCUE receives a 640 x 480 pixel gray-scale image (with 256 levels or 8 bits) stacked into a one-dimensional array with 307200 elements. These image measures are typical of CCD cameras. *Figure C-2* shows such an image.

In the image, two cues are on view: one white and one black. To the left, there is some wall texture that might produce some cue-like noisy structures. Also, a pattern consisting of three diagonal stripes has been introduced to evaluate false-positive rejection.

Figure C-2. Camera Image.

The first step that SCANCUE performs is a *quick and dirty* check of the transitions of pixel intensity of the image. The received image vector contains values in gray-scale from 0 – black – to 255 – white. A series of tolerances were determined experimentally to define when a transition between gray tones occurs. These tolerances are not absolute, because lighting conditions can vary from place to place.

SCANCUE searches for transitions in the gray-scale array that might account for Black-White-Black or White-Black-White that are roughly the same size within a specified tolerance. When SCANCUE detects a transition, it calculates the possible center of a cue. This process is shown in *Figure C-3*.

Once the routine has found the possible center, it takes three other passes: one in the vertical direction and two by the two diagonals. If SCANCUE finds, within a designated tolerance, the same White-Black-White or Black-White-Black ratio, it selects the structure to have at least cue-like properties. The algorithm then calculates the center, and the coordinates of the threshold of the change for each one of the scanning directions. *Figure C-4* shows these scanning directions.

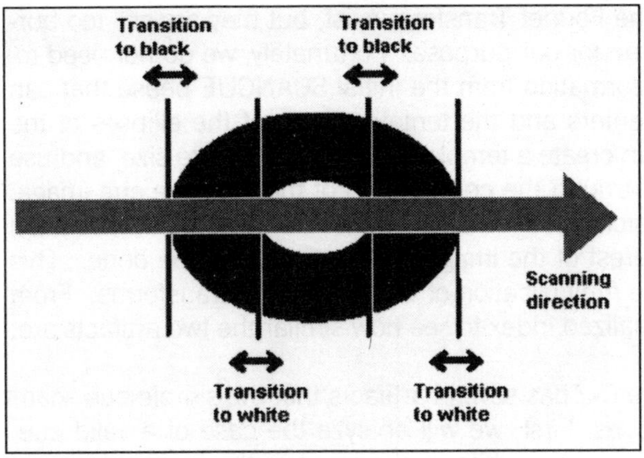

Figure C-3. Scanning First Pass.

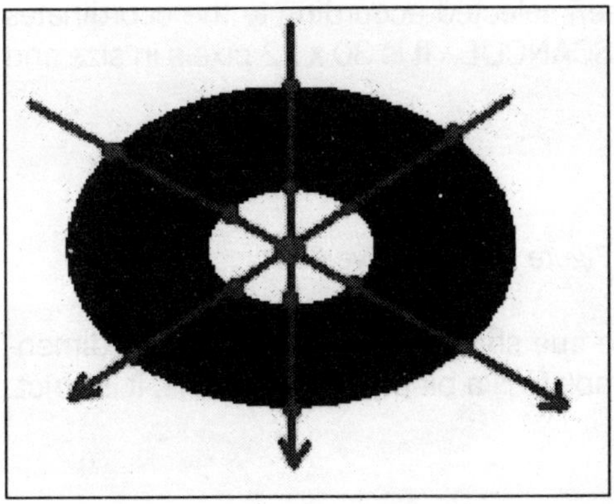

Figure C-4. Scanning Directions.

Using the threshold information, the routine can calculate the size of the inner and outer ellipses. This information will later be useful to measure how cue-like is the detected artifact.

The next phase of SCANCUE is to determine if the detected cue-like artifacts are in fact, cues. For that purpose, we employed an object recognition scheme called *pattern matching*. This technique consists in comparisons between an image and an image template or *mask*. For example, one could be searching for buildings in a picture taken from a plane, or for the outline of a body of water in a satellite image.

A standard way to do this comparison is to obtain the Fourier transform of both the image and the template, and then multiply the results. The resulting image will show very clearly where the match is closest, and a normalized index can be calculated to see how good the match is.

Fast algorithms to calculate the Fourier Transform exist, but they are still too burdensome in terms of computer time for our purpose. Fortunately, we do not need to, since there are some pieces of information from the initial SCANCUE phase that can be used to our advantage: the centers and the tentative sizes of the ellipses of the cues. With this information we can create a template of an approximate size, and use this mask to make a comparison around the central pixel of the tentative cue image. Instead of calculating the transforms, a calculation of the discrete convolution between the image portion of interest of the image and template can be done. This process is an approximation of the multiplication of the two Fourier transforms. From this result we can produce a normalized index to see how similar the two artifacts are.

The image displayed in *Figure C-2* has several artifacts that the simple cue scanning phase detects as possible cues. First, we will analyze the case of a valid cue. *Figure C-5* shows an enlarged view of the white cue present in the image shown in *Figure C-2*. The portion of the image has been selected according to the coordinates and the sizes detected in the earlier step by SCANCUE. It is 30 x 22 pixels in size and a 256 gray-scale image.

Figure C-5. Tentative Cue

We can generate a template of what the cue should look like Ousing the dimension data obtained in the first phase. This template is a binary image, that is, it is strictly in black and white. Observe *Figure C-6*.

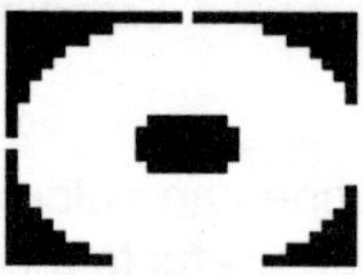

Figure C-6. Cue Template

Since the template is a binary image, it is convenient to convert the original image section into that format. For this purpose, the area of the template needs to be calculated. The template's area is the sum of its pixels, where a white pixel has a value of one and a black pixel has a value of zero. Then,

$$A_{Temp} = \sum_{i=0}^{N-1} \sum_{j=0}^{M-1} Tem_{ij} \qquad \text{(C.1)}$$

where the template is N x M pixels in size, its elements are Tem_{ij} and A_{Temp} is its area. In the case of the image in *Figure C-6*, A_{Temp} = 470 pixels.

A technique called thresholding can be now applied to the gray-scale image. The histogram of the image $H(j)$ is calculated, where the amplitude H is dependent on the gray-level frequency j. The cutoff frequency can be determined according to a certain criterion. Above that cutoff value, all frequencies are set as **1** (white) and below it all are set as **0** (black).

Before calculating the threshold value, white cues like the one shown in *Figure C-5* have to be inverted. That is, the values for the frequency at each point, Img_{ij} are replaced with 255 – Img_{ij}. That procedure is done to assess that the 'solid' area of the cue corresponds to the white part of the converted binary image. It also simplifies matters somewhat, since the same thresholding procedure can be applied to both types of cue after this transformation. *Figure C-7* shows the histogram of the inverse image of *Figure C-5*.

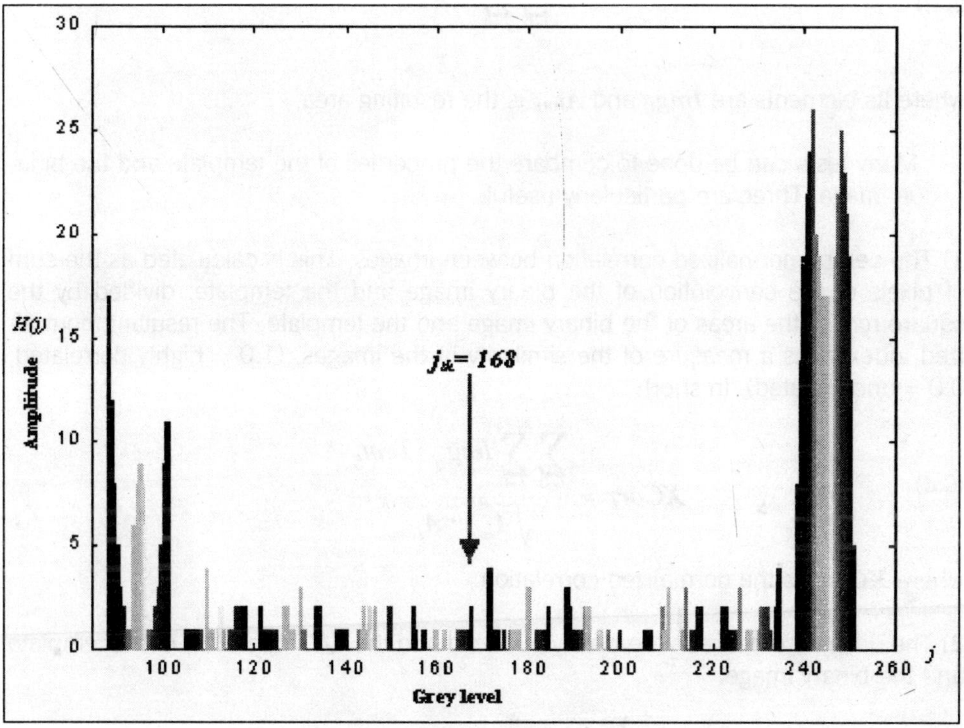

Figure C-7. Histogram of the Inverse of a Cue.

The threshold frequency is determined as

$$\sum_{j_{Thr}}^{255} H(j) \geq A_{Temp}, \qquad \text{(C.2)}$$

where j_{Thr} is the cutoff frequency. The sign is greater-than-or-equal in case the sum of the frequencies is not exactly the area of the template. For the current example, $j_{Thr} = 168$. Then, a binary image of the original cue can be constructed. It is shown in *Figure C-8*.

Figure C-8. Binary Image.

The area of the binary image can now be calculated,

(C.3)
$$A_{Img} = \sum_{i=0}^{N-1}\sum_{j=0}^{M-1} Img_{ij}$$

where its elements are $\boldsymbol{Img_{ij}}$ and $\boldsymbol{A_{Img}}$ is the resulting area.

Many tests can be done to compare the properties of the template and the binary cue image. Three are particularly useful:

1) The central normalized correlation between images. This is calculated as the sum of pixels of the convolution of the binary image and the template, divided by the square root of the areas of the binary image and the template. The resulting normalized index gives a measure of the similarity of the images. (1.0 = highly correlated, 0.0 = uncorrelated). In short:

(C.4)
$$XCorr = \frac{\sum_{i=0}^{N-1}\sum_{j=0}^{M-1} Img_{ij} \cdot Tem_{ij}}{\sqrt{A_{Img} \cdot A_{Temp}}}$$

where $\boldsymbol{XCorr}$, is the normalized correlation.

2) The difference between the binary cue area and the *AND* operation of the template and the binary image:

(C.5)
$$\Delta A = A_{Img} - (A_{Img} \cap A_{Temp}),$$

where $\boldsymbol{\Delta A}$ is the mentioned difference. The smaller this quantity is, the better the match between template and image.

3) The difference in pixels between an ellipse circumference calculated between the inner and outer ellipses of the template and an ellipse circumference calculated between the inner and outer ellipses of the binary image. This is a continuity check, to test if the cue ring of the binary image is roughly the same as the ring of the template. For convenience, this quantity is called ΔCir.

In the example, these are the calculated results for the valid cue:

$$XCorr = 0.950,$$
$$\Delta A = 24 \text{ pixels},$$
$$\Delta Cir = 0 \text{ pixels}.$$

Now, we will analyze two typical cases that produce false positives. The first one concerns texture on the wall that can have cue-like properties. This kind of random pattern can be present in bricks, plaster, etc. When the initial phase of SCANCUE parses the data in the image, it can confuse these patterns with very small cues, i.e. cues detected at a large distance from the camera. The image in *Figure C-2* has this noisy pattern on its left side, and the initial scanning detects at least 4 artifacts in that area that it considers tentative cues. For each of these structures, a template and binary image can be calculated. *Figure C-9* shows an example of a noisy false positive present in *Figure C-2*.

Figure C-9. Noisy False Positive.

The image fragment is shown at its actual size of 18 x 8 pixels. Note how the binary representation might be mistaken for a cue, a fact that is not readily obvious from the original gray-scale picture. The values calculated from the differences between the template and the binary images are:

$$XCorr = 0.70,$$
$$\Delta A = 27 \text{ pixels},$$
$$\Delta Cir = 11 \text{ pixels}.$$

Note how ΔCir detects that the ring is not continuous but $XCorr$ still shows that both structures are highly correlated. The case of ΔA is interesting: in binary images with mostly 'white' areas, this quantity provides little information about the properties of the gray-scale image, in particular with small fragments like the one shown in *Figure C-9*.

Another type of false positive that appears frequently is striped patterns or lines. This case is more common than might be apparent at first glance: sofas, furniture and wallpaper often have some sort of stripes. If these lines have the right angle, they might be able to fool the scanning routine into detecting a cue where there is none. *Figure C-2* has one such structure in its center-right part. Using the methodology, a template and a binary image can be calculated from a stripe-like artifact. One such case with dimensions 42 x 72 pixels is featured in *Figure C-10*.

Figure C-10. Striped False Positive.

The values calculated from the differences between the template and the binary images are:

$$XCorr = 0.780,$$
$$\Delta A = 135 \text{ pixels},$$
$$\Delta Cir = 16 \text{ pixels}.$$

In this case, ΔA gives the largest clue that the artifact of interest might not be a cue.

At this stage, we can define criteria for the confidence of cue detection. There will always be a risk of rejecting a true cue; but for our purpose, it is preferable to accept a false positive than to reject a false negative. It is possible to set a series of further filters based on current state estimates and *a priori* cue knowledge. But once an artifact is rejected as not being cue-like, the potential information is lost.

The following rules were set bearing those considerations in mind. The tolerances were obtained experimentally after extensive testing with different surfaces, textures, etc. First, we define the following parameters:

1) Threshold for $XCorr$ valid-cue criterion, THR_CORR_HIGH = 0.89
2) Threshold for ΔCir valid-cue criterion, THR_CYCLE_HIGH = 12
3) Threshold for ΔA valid-cue criterion, THR_AREA_HIGH = 80
4) Threshold for $XCorr$ false positive criterion, THR_CORR_LOW = 0.75
5) Threshold for ΔCir false positive criterion, THR_CYCLE_LOW = 12
6) Threshold for ΔA false positive criterion, THR_AREA_LOW = 30

Two tests are performed for each tentative cue. The first test judges whether the artifact is a cue. The second checks if the structure in question is a false positive. If the artifact does not pass either of these conditions, it is considered an ambiguous structure but it is not rejected.

An index of confidence, I_{conf}, can also be calculated. It gives a measure of how cue-like an ambiguous artifact is, one that has not been rejected outright as a false positive, but that does not show clear evidence of being a cue. The index has values between **0.0** (no confidence) to **1.0** (complete confidence).

A cue is considered valid then,

IF ($XCorr$ >=THR_CORR_HIGH *AND* $|\Delta Cir|$<=THR_CYCLE_HIGH) **AND** ($|\Delta A|$<=THR_AREA_HIGH)

The confidence index is calculated as:

$$I_{conf} = XCorr \times 0.8 + \left(1.0 - \frac{|\Delta Cir|}{THR\_CYCLE\_HIGH}\right) \times 0.1 + \left(1.0 - \frac{|\Delta A|}{50}\right) \times 0.1 \qquad \text{(C.6)}$$

An artifact is considered a false positive and rejected,

IF ($XCorr$ <THR_CORR_LOW) OR ($|\Delta Cir|$>THR_CYCLE_LOW) OR ($|\Delta A|$>THR_AREA_LOW).

In this case, $I_{conf} = 0.0$.

If the artifact does not fall in to either of these two categories, it is given the status of an undetermined structure and its index is calculated as:

$$\begin{aligned} I_{conf} = (XCorr \times 0.66 + \left(1.0 - \frac{|\Delta Cir|}{THR\_CYCLE\_LOW}\right) \times 0.166 \\ + \left(1.0 - \frac{|\Delta A|}{THR\_AREA\_LOW}\right) \times 0.166) \times 0.8 \end{aligned} \qquad \text{(C.7)}$$

This alternate definition of Iconf for indeterminate artifacts cannot get higher than 0.8.

Take for example the case shown in *Figure C-11*.

Figure C-11. Undetermined Artifact.

The 6 x 10 pixel artifact made of random texture resembles greatly a white cue. The values calculated from the differences between the template and the binary images are:

$XCorr = 0.823$,
$\Delta A = 8$ pixels,
$\Delta Cir = 2$ pixels.

These values do not pass either the validation or the rejection test and give the structure an index of confidence of 0.705.

The white cue present in *Figure C-2* passed the first, stricter condition (it was recognized as a valid cue) and has an I_{conf} = 0.960. The other two artifacts of *Figures C-10* and *C-11* were successfully characterized as false positives. For cases where the validity of the cue is not determined, false positives typically have values of I_{conf} of 0.60 or lower.

In conclusion, SCANCUE finds cues in an image by performing a two phase process: first, a simple and fast scan to detect cue-like structures; second, template matching and a series of tests to reject false positives. This scheme has been a valuable tool to reliably detect cues in a camera image. Most of the applications described in this book depend heavily on SCANCUE to obtain visual observations via cue detection.

APPENDIX D

DETERMINATION OF MATRIX JACOBIAN FOR LASER-SPOT CONVERGENCE

As discussed in Chapters 6-8, and 9, an enabling imaging tool is laser-spot recognition/coordinate-determination via image differencing together with application of a 2 x 2 matrix Jacobian to determine the approximate sensitivity of camera-space spot-center location to increments in the "pan" angle θ_p and increments in the "tilt" angle θ_t of the laser-pointer-bearing pan/tilt unit. Use of this Jacobian permits convergence of the spot center even with complex surface geometries in the "selection-camera space", as indicated in *Figure D-1*. With the laser spot converged onto the selection camera's juncture of interest, all participant CSM cameras can then register this same physical-surface location in the reference frames of their own camera spaces.

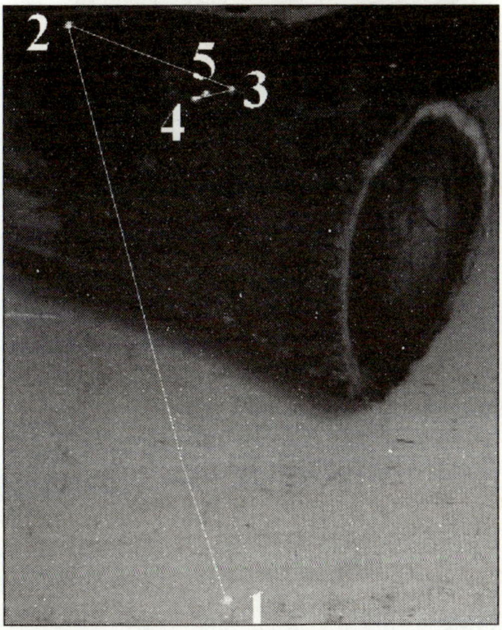

Figure D-1. Laser spot is controlled via approximate Jacobian as it moves sequentially from point 1 through to user-designated terminus, point 5.

We begin the discussion with the strategy for robustly locating the laser spot center in an image. This strategy is not only useful for locating rapidly and reliably the single laser spot used to produce convergence in *Figure D-1*; it is also useful for locating multiple laser spots in a single differenced image as indicated in *Figure D-2*. Even a visually complex surface such as the decorated wrapping paper in that figure can be handled with this device of applying image differencing – that is the subtraction, pixel by pixel, of image grayscale values with the laser pointer turned off from that with the laser pointer turned on.

The idea is to synchronize laser-pointer on/off with image acquisition. Subtraction of the laser-off image from the laser-on image results in a kind of isolation of the laser spot(s). As indicated in *Figure D-3*, most of the differenced image's pixels are near zero (compared to the typical 255-intensity-level maximum.) There is some "noise", or random-looking exceptions away from the spot itself, but pixels that comprise the spot are relatively very bright. The brightest laser-spot pixel in *Figure D-3* has an intensity of 93 whereas most of the pixels away from the spot's center are in single digits. Some could even be negative due to the differencing of noisy images.

Figure D-2. Most of the laser spots indicated in this image can be found quickly in camera space using image differencing, including spots that fall upon the visually complex wrapping paper. This is one instance where computer vision has an advantage over our natural vision. We have trouble seeing the spots on the wrapping paper. But computers can recall quantitatively information from both component images of the differenced image.

The exact strategy for writing software to locate the pixel coordinates of the spot center isn't critical. Many strategies are effective. It should be noted that the criterion for selecting a particular strategy isn't so much related to the question "where is the 'true' spot center?" (that is not defined) as it is getting all participant cameras to locate the spot center in physically consistent ways. In other words, it is important that the algorithm used to find the camera-space coordinates of the spot center has the attribute of finding center coordinates in all participant cameras that come as close as possible to representing the same surface juncture in physical space.

The simplest strategy - selecting the highest pixel in the region of the spot - yields camera-space coordinates of $\mathbf{x_c} = 5$, $\mathbf{y_c} = 4$, as can be verified in *Figure D-3*. These are the same coordinates that would be found by applying the template shown in

Figure D-3 above the grayscale image. This template is moved (virtually) such that its center lies immediately above any candidate spot-center pixel. A number is calculated for each of these positions which is the sum of the products of each nonzero template-grid element with the corresponding pixel intensity value. The template position with the highest such number is deemed the spot-center pixel.

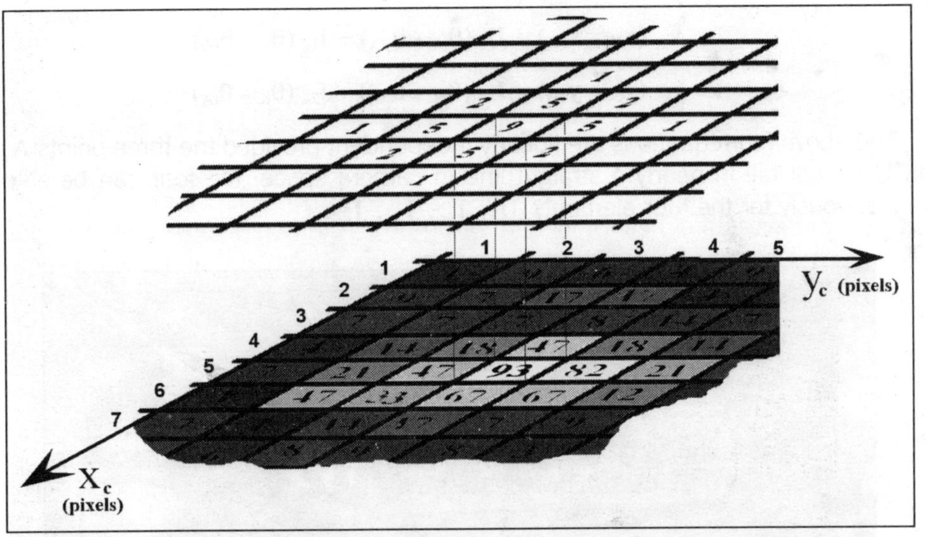

Figure D-3. Portion of a differenced image that includes a laser spot. Brighter pixels have higher intensity values. Template for best center location above.

It is generally a good idea to employ a template such as this. In fact, it is possible to achieve subpixel resolution in this way by allowing the template to interpolate between pixels. Programmers may enjoy identifying an effective strategy for achieving this.

Another interesting programming challenge, again one that admits several effective strategies, is using known pan and tilt angles together with corresponding detected laser spot locations in the selection-camera space - to find the four Jacobian elements J_{11}, J_{12}, J_{21}, J_{22}. The relationships of interest are:

$$\Delta x_c = J_{11}\,\Delta\theta_p + J_{12}\,\Delta\theta_t$$

$$\Delta y_c = J_{21}\,\Delta\theta_p + J_{22}\,\Delta\theta_t$$

Referring to *Figure D-4*, suppose we have available detected spot centers at points **A**, **B** and **C** together with pan and tilt angles corresponding with those camera-space spot readings. Considering the above general form together with the move from point **A** to point **B** we have:

$$(x_{cB} - x_{cA}) = J_{11} (\theta_{pB} - \theta_{pA}) + J_{12} (\theta_{tB} - \theta_{tA})$$

$$(y_{cB} - y_{cA}) = J_{21} (\theta_{pB} - \theta_{pA}) + J_{22} (\theta_{tB} - \theta_{tA})$$

Considering together points **A** and **C**, likewise, produces

$$(x_{cC} - x_{cA}) = J_{11} (\theta_{pC} - \theta_{pA}) + J_{12} (\theta_{tC} - \theta_{tA})$$

$$(y_{cC} - y_{cA}) = J_{21} (\theta_{pC} - \theta_{pA}) + J_{22} (\theta_{tC} - \theta_{tA})$$

The above four equations are linearly independent provided the three points **A**, **B** and **C** do not fall in nearly a straight line in camera space; the four can be solved simultaneously for the four elements J_{11}, J_{12}, J_{21}, J_{22}.

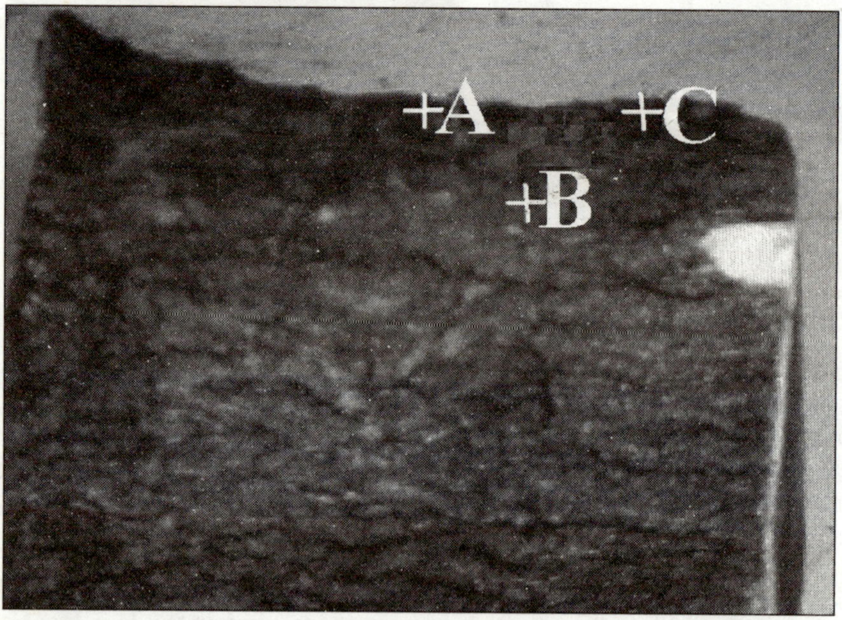

Figure D-4. How can we use the record of camera-space coordinates of **A**, **B** and **C**, together with the corresponding pan and tilt angles to estimate the four Jacobian elements?

There arises the question of the wisdom of choosing for this Jacobian-initialization three junctures **A**, **B** and **C** such as those indicated in *Figure D-4* that lie on the significantly curved and rough surface of the log. Interestingly, the physical convergence event over the same log of *Figure D-1* made use of a Jacobian based, as per the above, upon the laser spot falling on the flat surface of the table beneath which the log lay. The indirect convergence event over the log's surface reflects this fact.

In general, depending upon the character of the scene within which convergence will be pursued, it will be useful to apply redundant experimental information, often based upon the current convergence event to keep these Jacobian elements current. The techniques of *Appendix A* can be used to minimize a scalar function defined in terms of weighted residuals involving the Jacobian elements as the estimation parameters for this kind of overprescribed case.

INDEX

L

M

N

P

S

T

U

V

W

NOTES